AF718128

Range Ecology at Disequilibrium

New Models of Natural Variability and Pastoral Adaptation in African Savannas

edited by

Roy H. Behnke Jr.
Ian Scoones
and
Carol Kerven

Overseas Development Institute
International Institute for Environment and Development
Commonwealth Secretariat

Figure 1.3 was first published in J.E. Ellis and D.M. Swift, 'Stability of African pastoral ecosystems', *Journal of Range Management* Vol 41(6) p. 456, November 1988. Figure 12.2 was first published in G. Pickup, 'Event frequency and landscape stability on the floodplain systems of arid Central Australia', *Quaternary Science Review*, Vol 10, pp. 463-473, 1991, and is reprinted with permission of the publishers, Pergamon Press.

A CIP Publication data record may be obtained from the British Library.

ISBN 0 85003 195 8

SF 85.4 .A4 R35 1993

© Overseas Development Institute 1993

Published by the Overseas Development Institute,
Regent's College, Inner Circle, Regent's Park,
London NW1 4NS

in association with the

International Institute for Environment and Development,
3 Endsleigh Street London WC1H 0DD

and the

Commonwealth Secretariat, Marlborough House,
Pall Mall, London SW1Y 5HX.

All rights reserved. No part of this may be reproduced by any means, nor transmitted, nor stored electronically without the written permission of the publisher.

Typeset in PostScript at the Overseas Development Institute and Russell Press
Printed by the Russell Press Ltd, Nottingham

Contents

Tables

Figures

Plates

Preface and Acknowledgements

In 1987 the Commonwealth Heads of Government requested their Secretariat to examine the scope for Commonwealth action on soil erosion, desertification and related drought problems in Africa. One component of this overall programme was a project organised by the Food Production and Rural Development Division to address the potential for improved management of communal African rangelands. As a part of this effort, a technical meeting was held in Woburn, England, at the end of 1990 to explore the biological and ecological dimensions of African range management.

The meeting, which brought together 25 scientists with field experience throughout sub-Saharan Africa, examined a variety of new approaches to the study of African savanna ecology. It was clear from the meeting that there was considerable agreement among biological scientists on many fundamental issues, and that this consensus endorsed a thorough-going review of mainstream range management practice and theory. This book, which is based largely on papers presented at the 1990 meeting, is one part of that review.

The 1990 Woburn meeting was sponsored by the Commonwealth Secretariat and supported under their Commonwealth Fund for Technical Cooperation. Additional assistance was provided by the Environment Department of the World Bank, drawing on support from the Government of Norway for Sahelian environmental studies, to enable the conference papers to be collected and prepared for publication. Without the financial assistance of both the Commonwealth Secretariat and the World Bank this book could not have been produced.

Institutions, even very large ones, are ultimately represented by individuals, and in this respect we have been particularly fortunate. Brian Kerr of the Commonwealth Secretariat first raised in 1989 the possibility of a project on the management and sustainable use of communal rangelands in Africa. Since that date he has advanced the rangeland project through several phases with a variety of co-funders. This book is one manifestation of his determination, administrative skill, and persistent good humour. John English of the World Bank attended the 1990 meeting in Woburn, and his personal interest in the project played a significant role in securing financial support for this publication. In more intangible terms, his enthusiasm encouraged us to believe that this work had a potential audience among policy makers and development administrators who were sceptical about conventional rangeland management and willing to consider alternatives.

Individuals not represented by papers in this book who nevertheless attended the Woburn meeting as participants or commentators include Moses Awogbade, Ben Cousins, Kassim Farah, James Gambiza, Michel Grouzis, Cary Hendy, Maurice Keech, Walter Lusigi, Ruth Mace, William Payne, Jeremy Swift, Camilla Toulmin, and Richard White. Their contributions, either to debate in the meeting or in the form of written papers, were critical to the reformulation and sharpening of the arguments presented here, especially those in Chapter 1.

We would like to thank our respective institutions, the Overseas Development Institute (ODI) and the International Institute for Environment and Development (IIED) for material support, and Camilla Toulmin for moral support and practical advice throughout this project. We are also grateful to Margaret Cornell for editing the language and style of the manuscript, and to Jane Horsfield for her assistance in organising the 1990 meeting and for initiating the process of transforming meeting participants into book contributors.

Roy H. Behnke, Jr.
Ian Scoones
Carol Kerven

Contributors

N.O.J. Abel is a Lecturer at the School of Resource and Environmental Science, University of Canberra, Canberra, Australia.

Gerrit B. Bartels is a Range Management Specialist with Deutsche Gesellschaft für Technische Zusmmenarbeit (GTZ) in Abéche, Chad.

Roy H. Behnke Jr. is a Research Fellow at the Overseas Development Institute, London.

Yvan Biot is a Lecturer at the School of Development Studies, University of East Anglia, Norwich, UK.

Henk Breman is a Senior Scientist in the Agrosystems Research Department at the Directorate for Agricultural Research, Centre for Agrobiological Research (CABO-DLO), Wageningen, The Netherlands.

D. Layne Coppock is Assistant Professor in the Department of Range Science, Utah State University, Logan, Utah, USA.

Michael B. Coughenour is Senior Research Scientist at the Natural Resource Ecology Laboratory, Colorado State University, Fort Collins, Colorado, USA.

Lassine Diarra is an ecologist at the International Livestock Centre for Africa (ILCA), Bamako, Mali.

Peder N. de Leeuw is an ecologist at the International Livestock Centre for Africa (ILCA), Nairobi, Kenya.

James E. Ellis is the Director of the Center for Environment and Sustainable Agriculture, Winrock International, Morrilton, Arkansas, USA.

Pierre Hiernaux is an ecologist at the International Livestock Centre for Africa (ILCA), Bamako, Mali.

Brien E. Norton is Associate Professor, Department of Range Science, Utah State University, Logan, Utah, USA.

Gregory K. Perrier is International Programs Coordinator, College of Natural Resources, Utah State University, Logan, Utah, USA.

Geoff Pickup is the Head of the Commonwealth Scientific and Industrial Research Organisation's (CSIRO) Centre for Arid Zone Research in Alice Springs, Australia.

Nico de Ridder is a Lecturer at the Tropical Crops Sciences Section of the Agronomy Department, Agricultural University, Wageningen, The Netherlands.

Ian Scoones is a Research Associate of the Drylands and Sustainable Agriculture Programme at the International Institute for Environment and Development (IIED), London.

Mark Stafford Smith is a Senior Research Scientist with the Commonwealth Scientific and Industrial Research Organisation (CSIRO), Division of Wildlife and Ecology, at the Centre for Arid Zone Research in Alice Springs, Australia.

David M. Swift is Senior Research Scientist at the Natural Resource Ecology Laboratory and an Associate Professor in the Department of Range Science, Colorado State University, Fort Collins, Colorado, USA.

David Tapson is Manager of Development Planning at the Development Bank of Southern Africa, Headway Hill, Halfway House, Republic of South Africa.

John C. Tothill is an ecologist, now retired.

1
Rethinking Range Ecology: Implications for Rangeland Management in Africa*

Roy H. Behnke Jr. and Ian Scoones

Introduction

Few range management projects in dry Africa have had a discernible, positive, and permanent impact on the way communal rangeland is used. Most have failed to enlist the active cooperation of the pastoral communities they were supposed to serve. These failures reflect a variable combination of social, institutional and technical deficiencies in project and programme design. This overview examines one aspect of this complex problem: the limited appropriateness and validity of conventional range management theory in the African situation.

The third edition of Stoddart, Smith and Box's standard textbook *Range Management* opens with the observation that:

> In the more than 30 years since the appearance of the first edition of *Range Management*, there have been many changes... . Nevertheless, no new conceptual framework differentiates the field of range management now from then (1975:ix).

While this statement may have been true in 1975, it no longer holds. What were once anomalous individual field cases are now increasingly linked into an internally consistent, alternative theory of the functioning of savanna rangelands (Frost et al., 1986). In many instances this work calls into question conventional range management techniques and the theoretical assumptions which underpin these techniques.

The policy implications of the new ecological theories for Africa's predominately communal rangelands, managed by pastoralists, have been

---

* This overview is a slightly modified version of a paper previously published by The Commonwealth Secretariat (Behnke and Scoones, 1991) and by the World Bank as Environment Working Paper No. 53 (1992). The authors would like to thank Wolfgang Bayer, Henk Breman, Layne Coppock, John English, Peter Frost, James Gambiza, Cees de Haan, Dennis Herlocker, Brian Kerr, Ben Norton, Bill Payne, Gregory Perrier, Stephen Sandford, Martin Upton and Richard White for a close and critical reading of previous drafts of this report. We would especially like to thank Carol Kerven who commented upon the manuscript at every stage in its preparation. The views expressed here are those of the authors alone, as will be immediately apparent to a number of our commentators who raised substantial issues which we were not able to address within the confines of this introduction.

raised but only tentatively explored (Ellis and Swift, 1988). Likewise, the basic biological research which should inform policy making often is not readily accessible to the other parties interested in applied rangeland management, including administrators, social scientists and economists. The chapters in this book therefore review recent biological research on African rangelands and highlight its management implications for future donor and national government policy.

The preeminent management problem on communal African rangeland has been perceived for some considerable time, both by the public at large and by many rangeland professionals, as the control of rangeland degradation through the control of excessive livestock numbers. The scientific basis for this concern has been the concept of rangeland carrying capacity, defined and measured according to assumptions about the impact of herbivores on plant succession. This concept has provided the standard against which African rangelands are judged to be overstocked, inefficiently used, and ultimately degraded (Sandford, 1983a).

The contributions to this book pose a number of difficult questions regarding the precision with which carrying capacity can be estimated, current definitions of the concept and its relevance to certain dry African environments. They also critically examine the concept of rangeland degradation and propose techniques for its more appropriate assessment. Finally, these chapters contribute empirically to the debate by providing new data on the present condition of rangelands and livestock in a number of African countries.

In sum, it is argued here that the mainstream view of range science is fundamentally flawed in its application to certain rangeland ecologies and forms of pastoral production. If range management is to be of any use in these settings, conventional theories and recommended management practices require not minor adjustment but a thorough re-examination. This book provides an opportunity for just such a reassessment.

Carrying capacity and succession theory: the mainstream approach

The conventional notion of carrying capacity in range management rests on theories of plant succession, defined as the orderly and directional process whereby one association or community of plant species replaces another (Stoddart et al., 1975:156). Succession theory was initially developed at the turn of the century to explain variation in vegetation types in North America (Cowles, 1899; Clements, 1916). Research in range science from the 1920s to the 1940s transformed this theory into a practical, applied technique for the management of natural forage and grazing animals, that is, range management (Sampson, 1923).

Both succession theory and range management practice assumed that a single, persistent and characteristic vegetation, the climax, would dominate a particular site, depending on the soil and climate of that site. If this climax vegetation was disturbed, the vegetation could nonetheless return through a successional sequence to climax. An obvious example of disturbance and subsequent succession back to climax is provided by the clearing of a forest area for agriculture, the abandonment of the area, and the eventual reestablishment of forest through a predictable sequence of intermediate vegetational stages.

Range management adapted these ideas to grazing systems. It was assumed that the effects on vegetation of grazing paralleled, in a less dramatic way, the effects of clearing fields for crop agriculture. That is, grazing pushed the successional sequence back to some form of sub-climax. The task for the range manager was to balance grazing pressure against the natural regenerative power of the plants, thereby maintaining a stable sub-climax which yielded a steady and profitable flow of animal products. The concept of carrying capacity was important because it marked the stocking density at which this balance could be achieved.

Pushed beyond the threshold of carrying capacity, the balance between grazing pressure and the inherent regenerative powers of the range was destroyed, and the condition of the range progressively deteriorated. This deterioration was reflected in a process of regression back through the successional sequence. The theoretical relation between poor range condition and an early stage in a successional sequence is diagrammatically expressed in Fig. 1.1.

In practical terms, experienced range managers found that they were often able to estimate range condition by reference to plant species which were particularly sensitive to the effects of grazing. These indicator species either increased, decreased or invaded a range depending on the intensity of grazing pressure, and thereby provided a convenient measure of the extent to which grazing had altered and was continuing to alter the climax vegetation. This botanical approach to the assessment of range deterioration was defended on the grounds that vegetation change preceded both reduced livestock production and increased levels of soil loss, and therefore served as a valuable 'early warning' of declines in other parts of the rangeland system (Stoddart et al., 1975:267).

Carrying capacity: ecological or economic

A different approach to the definition of carrying capacity has been developed by wildlife population biologists, in response to the practical problems of managing parks, their vegetation and their wild herbivore populations. Range ecologists have much to learn from these allied professions in developing a definition of carrying capacity which is appropriate to the management of communal rangelands used by African pastoralists. If extended to include the study of pastoral production on communal ranges, this approach to carrying capacity also demands a fundamental reassessment of the extent to which heavy stocking rates in pastoral areas constitute overgrazing.

Fig. 1.2, originally presented by Caughley (1979) and elaborated by Bell (1985), provides a schematic overview of the relationship between plant and wild herbivore populations at alternative stocking densities. The top curve in Fig. 1.2, called the zero isocline of vegetation, marks all technically feasible combinations of plant and animal densities in a hypothetical grazing system. At the far right end of the horizontal axis, the curve depicts the situation

Figure 1.1: Relationship between range condition and degree of retrogression from climax conditions

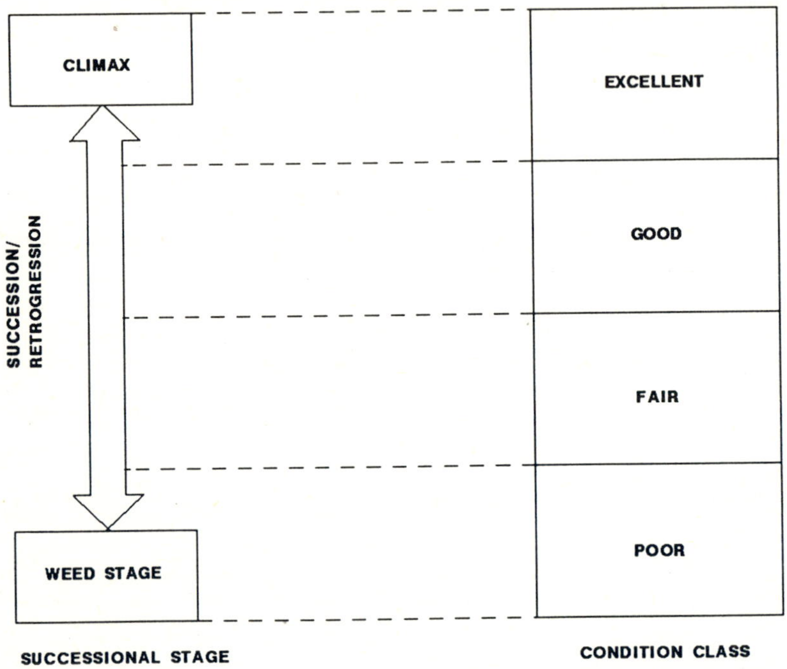

Source: Modified from Stoddart, Smith and Box (1975)

which prevails when there is a small animal population and a large standing crop of plants. As the animal population increases, the edible plant biomass declines. In an undisturbed grazing system, the increase in animal numbers will eventually be checked by the declining availability of natural forage. This will occur when the production of forage equals the rate of its consumption by animals, and the livestock population ceases to grow because limited feed supplies produce death rates equal to birth rates. At this point there is no surplus production either of individuals or biomass. This point of equilibrium, routinely designated 'K' in the ecological literature, is termed ecological carrying capacity in Fig. 1.2. At ecological carrying capacity, livestock may be plentiful but they will not be in particularly good condition; neither will the vegetation be as dense nor will the plant communities necessarily be composed of the same species as they would be in the absence of animals (Caughley, 1979; Bell, 1985).

If managers want denser vegetation or healthier animals, then they must maintain fewer animals. This can be done either by hunting, in the case of wild herbivores, or by culling, in the case of domestic stock. The offtake curve in Fig. 1.2 indicates the different offtake levels managers must maintain in order to support combinations of plant and animal densities other than those occurring at ecological carrying capacity. Initially the offtake curve rises from zero at very low stocking rates and increases with the increasing size of the herbivore population. The sustainable offtake rate – determined by multiplying the total animal population by the excess of the birth over the death rate – is highest at the stocking density at which the animal population is growing most rapidly. This point of maximum sustained yield usually lies at about half to two thirds of the stocking density at ecological carrying capacity, a stocking density which Caughley has termed 'economic carrying capacity' (1979). As the animal population grows beyond economic carrying capacity the offtake rate begins to fall and ultimately returns to zero as increasingly high rates of mortality and falling birthrates obviate both the need and opportunity for offtake to maintain stable animal populations.

Depending on the economic and aesthetic environment in which they are operating, wildlife managers have been called upon to maintain many of the different combinations of plant and animal densities illustrated in Fig. 1.2. Through their control over hunting quotas, they frequently have had the capacity to do so.

Figure 1.2: The relationship between plant and animal populations in a grazing system

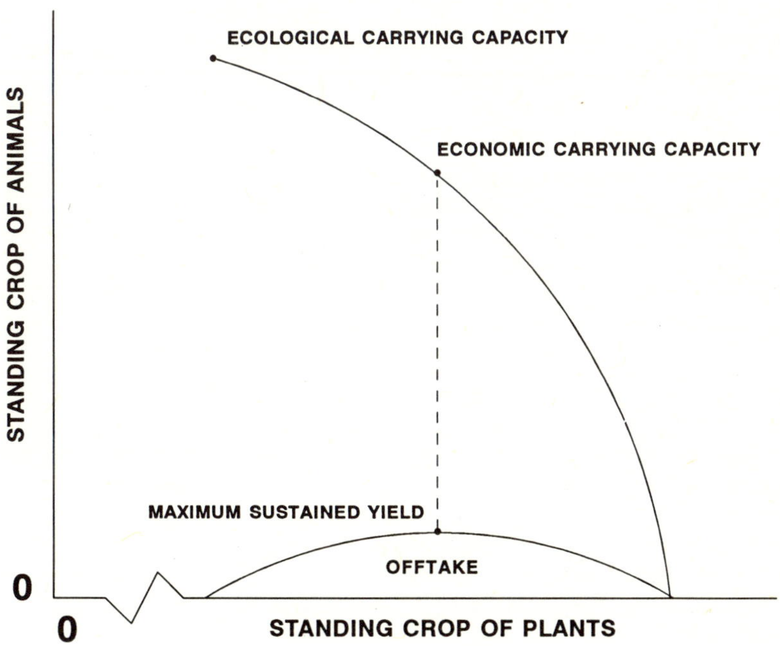

Source: Adapted from Caughley (1979) and Bell (1985)

As an illustration of the management options open to wildlife ecologists, let us consider a park which is financially sustained by a tourist industry based on game viewing. In this case the manager will require a relatively dense population of animals which will increase the probability that the individual tourist will actually confront the animals he has come to see. In this instance the park manager may desire a high animal population well above economic carrying capacity, a density which might be termed 'camera carrying capacity'. An unavoidable, but potentially attractive, by-product of these high stocking rates might be a thinning of the vegetative cover which could interfere with the sighting of game. On the other hand, a park might be operated to produce maximum kilograms of game meat for sale. In this instance the manager will require that density of animals which provides the maximum sustained yield in terms of meat output, or economic carrying capacity as defined by Caughley. An unavoidable by-product of this management system will be fewer animals and more vegetation, relative to a park managed for game viewing. Still other animal-plant population balances might be required in parks managed to produce trophy specimens, or to preserve particular plant communities sensitive to grazing pressure.

And which of these three park management systems is the correct one? All are technically feasible and all are economically profitable, under certain conditions. And each is associated with a distinctive density of animals. From the vantage point of wildlife management all these management systems are ecologically and scientifically defensible, although their relative financial and aesthetic merits might be hotly contested. Implied in this position is the conclusion that there is no single biologically optimal carrying capacity which can be defined independently of the different management objectives associated with different forms of animal exploitation.

> We conclude, therefore, that the only embracing definition of carrying capacity is: 'That density of animals and plants that allows the manager to get what he wants out of the system'. Thus, any specific definition of carrying capacity must be expressed in relation to a particular objective, and it must be defined very precisely since there are no 'natural' stability points in such interactive systems that act as foci for self-defining concepts (Bell, 1985:153).

Given this perspective, it makes little sense to speak about overgrazing or understocking unless managers also specify the kind of management system they wish to institute and frame their assessment in terms of the appropriate stocking density for that system. Contrary to the presumptions of mainstream range science, there exist for wildlife managers no 'objective' biological criteria which will permit the specification of carrying capacity without prior reference to the goals and objectives of managers.

And if carrying capacity must be defined relative to economic objectives for wildlife management, why should the concept be treated any differently when applied to alternative forms of domestic livestock production on natural forage? In the pastoral as in the wildlife setting, there would appear to be different stocking densities associated with and appropriate to different forms of pastoral production. For example, if there exists consumer demand for high-grade meat, some ranchers may find it profitable to sell relatively few animals in excellent condition raised on a relatively abundant forage supply. These ranchers will need to hold their stocking densities well below economic carrying capacity as defined by Caughley, and will have to accept slaughter offtake rates below maximum sustainable yield expressed in terms of kilograms of harvested meat. Alternatively, ranchers may be producing for a market in which meat is sold ungraded by weight, as is presently the case in Kenya. Ranchers operating in this marketing environment will, like their counterparts producing game meat sold by weight, seek to maintain stocking densities close to Caughley's economic carrying capacity[1].

Finally, there is the case of subsistence-oriented pastoralism as well as other forms of livestock husbandry (commercial dairy and fibre production) which seek to harvest animal output in the form of live-animal products such as milk, blood, traction power and transport. Offtake for these producers does not require animal slaughter and they can, therefore, profitably exploit a large standing crop of animals (Payne, 1990). At some cost in terms of the output, health and viability of individual animals, these producers may be capable of maintaining high levels of aggregate output at stocking densities approaching ecological carrying capacity. Natural mortality in such heavily stocked systems may be high, but for the pastoralist it is not the unmitigated disaster it would be for commercial ranchers since animals can be slaughtered in anticipation of death and, in some cases, a certain percentage of carcasses may be retrieved and consumed after death.

The relationships depicted in Fig. 1.2 are simplified and cannot predict real plant and animal interactions in most grazing systems. Subsequent sections of this introduction elaborate on many of the additional factors which must be considered in evaluating the effects of grazing pressure on rangeland resources in different situations. What Fig. 1.2 does provide is a logical structure for distinguishing between the ecological and economic aspects of rangeland assessment.

Mainstream range management has sought to develop the biological science of rangeland use in order to address the practical needs of producers. Due to the historical association of range management with producers on beef ranches, many of the standard botanical indicators used to assess 'carrying capacity' (increasers, decreasers, perennial:annual ratios, bush 'encroachment' etc) have actually been implicitly derived in order to assess *economic* carrying capacity levels for beef ranching systems. Here we have one explanation of how livestock numbers in some parts of Africa have continued to grow, in some instances for four or five decades, beyond the purported limits of 'carrying capacity'. What was being estimated by the techniques of range

---

1. Because there are significant variable costs associated with holding domesticated stock, economically optimal stocking densities for commercial ranchers will always lie below the stocking density which produces maximum sustainable yield per hectare (Workman, 1986; Wilson and Macleod, 1991; Jarvis, 1984; Carew, 1976.)

management, it would appear, were not ecological but economic carrying capacity levels, and moreover, economic carrying capacity levels for kinds of production systems which did not exist in the areas being assessed. In Zimbabwe, for example, official recommended stocking rates relate to 'economic' carrying capacity for commercial beef production, and are a half to a third of estimated ecological carrying capacity and well below long-term stocking densities (Scoones, chapter in this book)

Grazing systems not at equilibrium

The erratic and variable rainfall in many pastoral areas of Africa poses a further fundamental challenge to standard conceptions of carrying capacity. Any notion of carrying capacity – be it ecological or economic – is predicated on the notion that herbivore numbers are controlled through the availability of forage and that the availability of forage is controlled by animal numbers, a pattern of negative feedback which eventually produces a stable equilibrium between animal and plant populations.

This pattern of interaction between plants and herbivores presumes, in turn, that conditions for plant growth are relatively constant. If physical factors such as rainfall and temperature fluctuate widely, it is likely that these non-biological variables will have a greater impact on plant growth than marginal changes in grazing pressure caused by different stocking densities. Moreover, unavailability of forage in bad years may depress livestock populations to the point where the impact of their grazing on the vegetation is minimal in most years. Thus, in these fluctuating climates, rainfall, not forage availability, may ultimately be the variable which limits herbivore population growth.

If disturbances are intermittent, it may be useful to analyze a grazing system as if it were at equilibrium, and to treat outside perturbations as 'noise' which confuses and obscures an underlying equilibrium pattern. On the other hand, if disturbance is frequent, random 'noise' so dominates events that it is more useful to think of the 'noise' itself as the system. Noisy or event-driven grazing systems require a different approach to and understanding of carrying capacity, which we must now examine.

Fig. 1.3, based on Ellis and Swift (1988), illustrates plant-livestock interactions under the influence of frequent drought perturbations in a fluctuating climate – that of Turkana, Kenya. The axis labels in Fig. 1.3 are identical to those in Fig. 1.2. What has changed is the presumed level of stability in the grazing system. As a result, the inverse relationship between plant and animal populations which characterised Fig. 1.2 has been replaced by a more complicated pattern.

The points on the far right of Fig. 1.3 chart a process of both plant and animal population expansion under favourable rainfall conditions for that particular environment. The points to the left of the figure represent the contraction of both populations under drought conditions of varying degrees of severity. Single year droughts constitute a minor and very temporary setback for the animal population, and a somewhat greater but nonetheless temporary setback for the plants, while multi-year droughts precipitate population crashes of both plants and animals. In this system, livestock

Figure 1.3: Turkana plant-livestock interactions under the influence of frequent drought perturbations

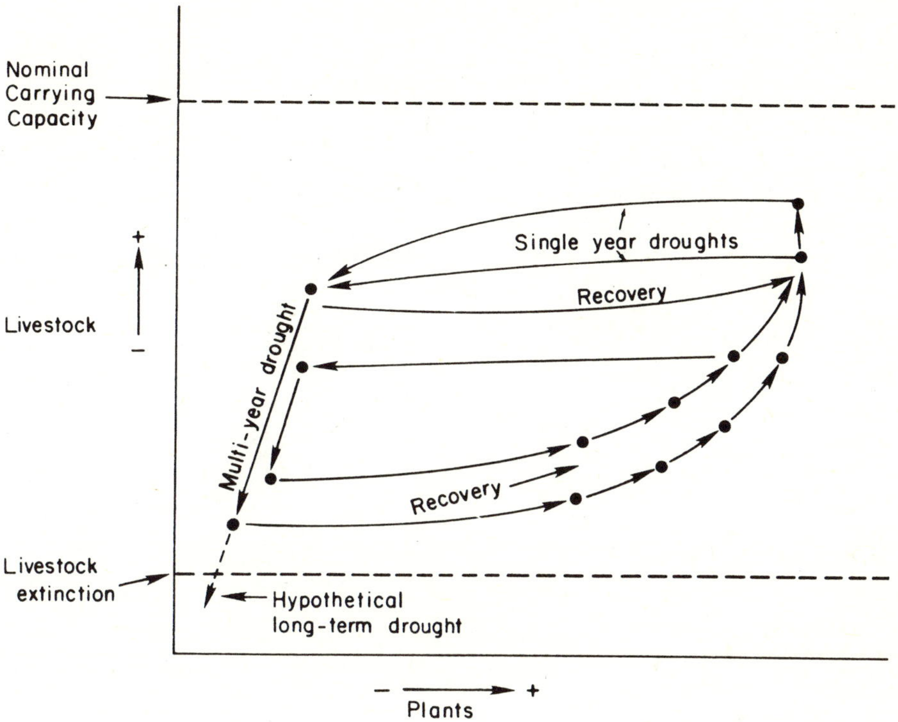

Source: Reproduced from Ellis and Swift (1988)

populations may decline because of a lack of fodder, but fodder is scarce because there is too little rain rather than too many animals. Moreover, major droughts are frequent enough and herd recovery is slow enough that livestock numbers are never given an opportunity to approach ecological carrying capacity. In sum, the condition of this grazing system at any particular time is determined more by the chance occurrence of non-biological events than by interaction between the biological components of the system itself (Ellis and Swift, 1988).

Why this should be so is illustrated in Fig. 1.4, which summarises the differences and underlying similarities between equilibrium and non-equilibrium grazing systems. In contrast to Fig. 1.2. (the equilibrium situation), Fig. 1.4. is based on a series of alternative vegetation isoclines corresponding to different annual rainfall levels, rather than one such level. The plant and animal populations which could theoretically be supported during an extended period of mean rainfall are presented in the middle curve. On either side of the mean isocline are additional curves depicting potential plant and animal

Figure 1.4: Schematic representation of plant-livestock interactions under the influence of frequent drought perturbations

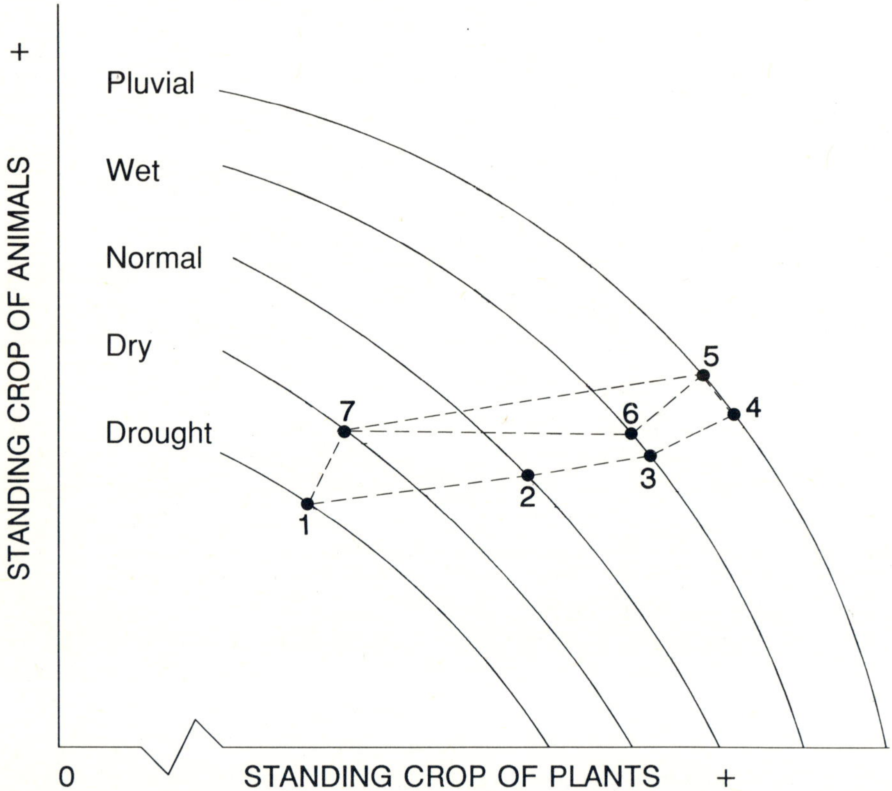

populations in periods of above or below average rainfall. These additional curves reflect the diminished importance of mean rainfall and mean production values for an understanding of a system dominated by variability. For illustrative simplicity, only four additional curves are given, representing pluvial, wet, dry and drought years. In reality, variable amounts and timing of rainfall generate an almost limitless number of such additional curves.

Superimposed upon this series of isoclines are seven data points which, in much simplified form, depict the pattern of livestock and plant population response to variable rainfall in Turkana. As in Ellis and Swift's initial diagram (Fig. 1.3), movement along the pathway 5—7—1 represents the onset of a major multi-year drought; pathway 1—2—3—4—5 represents recovery from such a drought; and the circuit 5—7—6—5 depicts the impact of and recovery from a single year drought. These alternative 'pathways' (or sequential combinations of plant and animal populations) are not confined to movement along one isocline, as they would be in the equilibrium situation. They instead reflect movement across a number of hypothetical isoclines, as plant and animal populations respond differently to short or long periods of deviation from mean rainfall levels, and respond at different rates.

Figures 1.2, 1.3 and 1.4 summarise quite different approaches to the understanding of rangeland ecology. In equilibrium grazing systems of the kind depicted in Fig. 1.2, the physical conditions supporting plant growth are relatively unvarying, consumption by herbivores controls plant biomass, and the availability of feed ultimately regulates the growth of the herbivore population. In Figs. 1.3 and 1.4 depicting non-equilibrium grazing systems, the physical conditions supporting plant growth vary widely and consumption by herbivores does not control plant biomass because the animal population is itself held in check by the same physical factors which control the vegetation. Grazing pressure may cause changes in vegetation, but the effects are complex and intermittent, as is discussed below.

Vegetation change in an episodic environment

Thus far the discussion has focused on the relationship between plant and animal biomass in a grazing system. But range managers are not interested solely in the quantity of forage available for livestock, but in its quality and, hence, in the species composition of that forage. Grazing systems not at equilibrium present peculiar problems for the analysis and management of compositional changes in rangeland vegetation.

Irreversible, sudden or unpredictable changes in vegetation are difficult to reconcile with conventional notions of range succession as an incremental response to grazing pressure (Bartolome, 1984). Citing extensive evidence of such anomalies, Westoby et al. have argued that standard successional models cannot account for observed patterns of vegetation change in rangelands not at equilibrium, and have offered an alternative 'state-and-transition' model to account for these changes.

In this model, no attempt is made to array various vegetation states along a single successional pathway. Instead, the vegetation in a particular area is described 'by means of catalogues of alternative states and catalogues of possible transitions between states' (Westoby et al., 1989:266). A range may move from one state into a number of different states, or return to its original state along a transitional pathway, and due to factors different from those which caused the initial change.

Different combinations of factors, of which grazing pressure is but one element, may be required to cause an alteration in state, and the effects of a particular stocking density will be unpredictable unless all these factors are known. Because other factors vary widely, effectively managing arid rangelands is not a matter of adhering to a single, conservative stocking rate which will apply in all circumstances. Rangeland management is, instead, a game of calculating probabilities 'the object of which is to seize opportunities and to evade hazards, so far as possible', what Westoby et al. (1988:266) call 'opportunistic management'.

The implications of opportunistic management for formal livestock development policy in dry Africa will be discussed in the closing section of this introduction. Opportunism is not, however, new to Africa's pastoralists; it provides the rationale behind one of the most characteristic of their husbandry techniques – migratory stock keeping.

The ecological determinants of livestock movements

Livestock movement is likely to play a very different role in equilibrium and non-equilibrium grazing systems. If a herd is confined to one place, livestock numbers, viability and productivity are limited by the scarcest resource in the scarcest season in that place. These limits to settled livestock husbandry will apply, both in an equilibrium grazing system of the kind depicted in Fig. 1.2. or in a non-equilibrium system of the sort depicted in Figs. 1.3. and 1.4. But the costs of immobility will be slight in equilibrium systems where conditions are constant, and high in non-equilibrium systems where one particularly unfavourable period can limit production irrespective of the abundance of resources in other periods.

Mainstream range management techniques are ideally suited to addressing the needs of settled forms of animal husbandry operating under equilibrium conditions (exemplified by fenced ranches in temperate climates). Essentially, these techniques attempt to dampen seasonal and inter-annual resource fluctuations within a delimited rangeland area. Conservative stocking rates, for example, are designed to provide a prudent rancher with a 'buffer' of surplus forage in unusually poor years; fencing or the placement of water points is used to promote uniform patterns of grazing and efficient forage consumption, while cultivated pastures are intended to offset insufficient forage production on natural pastures in certain seasons, etc. These techniques are useful in equilibrium grazing systems in which range productivity is both reliable and susceptible to some degree of management control.

Non-equilibrium grazing systems present a different kind of management problem. The costs of a sedentary production strategy are likely to be much higher in non-equilibrium settings because of the wide, unpredictable, and largely uncontrollable swings in productivity which characterise these environments. Here effective management is more a process of responding flexibly to stress rather than preventing it, and movement provides a means of circumventing stress under certain ecological conditions.

The advantages of herd mobility are illustrated schematically in Table 1.1. which depicts a mixed settled and migratory grazing system consisting of three ecological zones used over three seasons. The values in the table represent the potential number of livestock which could be sustained in each zone by season, assuming wide seasonal variation in zonal carrying capacities. In this hypothetical system, the number of sedentary livestock which can be maintained permanently in any ecological zone is 100, the carrying capacity of each of the zones during their seasonal period of most restricted resource availability. The total sedentary livestock population which can be supported within the region is 300, the sum of the lowest carrying capacities of the three ecological zones.

Mobile livestock production would increase the total regional livestock carrying capacity to 1,000. Permutations of eight different migratory regimes

Table 1.1: Settled and migratory stock levels in a seasonally variable environment

Regional ecological zones	*Seasons* *Wet*	*Transitional*	*Dry*	*Settled stock numbers*
A	1,000 →	100	→ 200	100
B	200 ⇢	→ 1,000 →	100	100
C	100	⇢ 200 ⇢	→ 700	100
Totals			1,000[a]	300[b]

Notes: [a] Total regional livestock population including migratory and settled stock
[b] Total regional sedentary livestock population

could be employed by individual herders to sustain this increase (moving sequentially through the wet, transitional and dry seasons, these regimes are ABC, ACB, ABA, ACC, BCA, BBA, BBC, BCC). A simple migratory pattern combining two such regimes is illustrated in the table. Assuming 100 permanently settled animals in each zone, 600 additional migrant animals could be sustained on the migratory cycle A (wet) → B (transitional) → C (dry) → A (wet), indicated by the solid arrow. A further 100 migrant animals could be sustained by a B (wet) → C (transitional) → D (dry) → B (wet) pattern of movement, indicated by the broken arrow. These increases are possible because migratory stock numbers are determined by the scarcest resource period in the region as a whole (the dry season in our hypothetical example), rather than the sum of each such period for individual ecological zones.

Although not intended to illustrate any concrete situation, Table 1.1. suggests a transhumant pattern of cyclical herd movement based on predictable environmental fluctuations. Although rigidly simplified for illustrative purposes, this case involves eight distinct migratory regimes which could be variously recombined depending on the number of animals following each regime. Field studies of pastoral transhumant cycles confirm the potential complexity of these systems (Fry and McCabe, 1986; Dyson-Hudson, 1972; Behnke and Kerven, 1984).

This complexity is magnified by the effect of fortuitous environmental fluctuations, analyzed by Sandford (1983a:33-36). The logic of Sandford's analysis is similar to that in Table 1.1, but is based on annual rather than seasonal carrying capacity figures for grazing areas within a region. A simplified version of Sandford's analysis is recalculated in Table 1.2.

In Table 1.2, area A is a relatively high production zone, B is medium, and C is a relatively unproductive zone, measured in terms of each area's three-year mean and single-year maximum and minimum carrying capacity. What matters in this case, however, are not the permanent ecological differences

Table 1.2: Settled and migratory stock levels in an unpredictably variable environment

Regional ecological zones	*Three year mean carrying capacity*	*Years* 1	2	3	*Settled stock numbers*
A	633	1,000	400	500	400
B	433	300	200	800	200
C	333	100	700	200	100
Totals	1,400[a]		1,300[b]		700[c]

Notes: [a] Regional mean carrying capacity
[b] Total regional livestock population including migratory and settled stock
[c] Total regional sedentary livestock population

among the areas, but transient differences in forage production resulting from the erratic distribution of rainfall in particular years.

If livestock populations must be held in each area separately, the total sustainable regional livestock population over the three-year period is 700 head, the sum of the carrying capacities for all areas in their worst year. On the other hand, if we presume that animals can move freely between areas in response to exceptionally high or low rainfall, a total regional livestock population of 1,300 can be maintained. This higher value reflects the combined total carrying capacity of all three areas in the worst rainfall year for the region as a whole, year two.

Table 1.2. and Sandford's more detailed calculations illustrate the benefits of opportunistic stock movement in response to unpredictable rainfall fluctuations which are spatially and temporally random, and are suggestive of contemporary patterns of herd movement in the communal areas of Botswana and Zimbabwe, or Kenyan Maasailand. In these cases long-distance livestock movement is predominantly a contingent response to unpredictable but localised rainfall deficits, disease outbreaks, borehole breakdowns or range fires. In contrast to pastoralists engaged in seasonal transhumant movement, herders in these areas do not follow regular migratory routes. They instead maintain access rights to safe havens or fallback areas which will carry their herds through temporary crises in their home area.

Whether movement is regular and seasonal, contingent, or a combination of contingency and regularity, the producer's strategy within non-equilibrium systems is to move livestock sequentially across a series of environments each of which reaches peak carrying capacity in a different time period. Mobile herds can then move from zone to zone, region to region, avoiding resource-scarce periods and exploiting optimal periods in each area they use. In this way mobile livestock producers can maintain within a wide geographic region a total livestock population and levels of productivity in excess of that which could be sustained, all else being equal, by several separate herds confined to their individual areas. The prevalence of herd mobility as a husbandry strategy is symptomatic of the general approach to livestock management in non-equilibrium environments. Herd management must aim at responding to alternate periods of high and low productivity, with an emphasis on exploiting environmental heterogeneity rather than attempting to manipulate the environment to maximise stability and uniformity. The closing sections of this document will discuss the implications of this 'opportunistic' style of herd management for the design of formal livestock development projects and programmes.

Responses to spatial and temporal variation: three cases from pastoral Africa

Three chapters in this book – which describe pastoral systems in Kenya, Ethiopia and Zimbabwe – explore the distinction between equilibrium and non-equilibrium grazing systems. Each of these chapters also emphasises the role of opportunistic movement – on a daily, seasonal, yearly and generational scale – in the maintenance of these grazing systems. Probably the most exhaustively studied non-equilibrium or 'event-driven' grazing system in pastoral Africa is that of the Turkana of northwestern Kenya, described by J. Ellis and his co-workers (reference to this work can be found in Coughenour et al., 1985; Coppock et al., 1986; Ellis et al., 1987; Ellis and Swift, 1988 and Ellis et al., chapter in this book). Ellis and his colleagues found that in central Turkana, rainfall levels affected all aspects of the production system, and were highly erratic. Drought had occurred about 13 times in the last 50 years, and serious multi-year drought had occurred four times over this time period (Ellis et al,. 1987). Livestock losses due to drought could cut herd sizes in half, but there was little evidence that rates of loss were closely related to stocking rates. Basically, animals begin to starve, or at best hold their own, during the dry season. If the dry season was prolonged by drought, termites and the loss of vegetation to wind, sun and decomposition removed dry forage even if it was not consumed by livestock. With the exception of certain localities which sustained very high stocking densities, how many animals made it through a drought was determined more by the length of the dry period than by the number of animals which existed before the dry period began.

Ellis and Swift (1988) broaden the scope of this analysis to include arid grazing systems outside Turkana. They examine long-term rainfall patterns from a number of arid regions in Africa and argue that many of these environments experience massive and unpredictable fluctuations in rainfall similar to those in Turkana. Given these climatic patterns, non-equilibrium, event-driven grazing systems may prevail on many of the most arid rangelands of the continent.

These conclusions are qualified in Coppock's study of the Borana rangelands of southern Ethiopia, in this book. Rainfall is higher and more reliable in Borana than in Turkana, and severe droughts occur at less frequent, 20-year intervals. Coppock argues that in this more stable environment, pastoralists and their livestock are important agents of vegetation change. Periodic droughts may make interpretation of the situation more difficult, but the fundamental pattern is one of equilibrium, and equilibrium concepts such as carrying capacity are therefore analytically useful in the context of these environments.

In Borana, however, the pattern of grazing-induced vegetation change is complex both spatially and over the long-term. Elaborating on the work of J.C. Billé and others, Coppock hypothesises a process of bush encroachment under heavy grazing pressure which depletes soil nutrients and increases the competitive advantage of shrubs over perennial grasses. The replacement of grasses by woody shrubs is followed by the abandonment of the site by pastoralists. In the absence of heavy grazing pressure, few new shrubs are established, while those which already exist grow to maturity. Soil nutrients are slowly replenished by leaf litter, and grasses are gradually reestablished as fires thin out the trees. In a cycle that can take from 60 to 100 years to complete, the pastoralists recolonise the site which once again has a fertile soil and supports a mixed grass and tree savanna.

Although the composition of the vegetation at any particular site is unstable, the overall grazing system in Borana may be remarkably persistent, as pastoralists cycle through a number of different sites. This pattern of land use raises both theoretical questions regarding the nature of degradation and practical questions regarding the appropriateness of measures to control it in Borana. Within mainstream rangeland management, bush encroachment, the loss of soil and soil nutrients, and declining livestock productivity indicated by the abandonment of sites by pastoralists would qualify unequivocally as rangeland degradation. But as described for the Borana case, bush encroachment is part of a potentially sustainable pattern of rangeland use built around spatial flexibility by pastoral producers. Efforts to control bush encroachment and stabilise productivity at a particular site would forestall the very processes which eventually rejuvenate site productivity and provide the basis for continued rangeland productivity on a regional scale. What is critical to the maintenance of the larger system are human and livestock populations which are low enough to permit sufficient 'fallowing' between the reoccupation of individual sites.

Like Coppock, Scoones (chapter in this book) is concerned with disentangling the relative importance of equilibrium and non-equilibrium factors in shaping a grazing system, in this case a communal area in Zimbabwe. Unlike Coppock, Scoones focuses his analysis on the dynamics of the livestock populations rather than the state of the vegetation. He does this by asking what controls the growth of the livestock population – particular historical and episodic events such as droughts, or continuous, systemic factors such as the size of the cattle herd itself.

Using sixty years of livestock population data from southern Zimbabwe, Scoones concludes that in a run of relatively good rainfall years cattle populations do approach a ceiling set by ecological carrying capacity. As stocking densities increase, birth rates decline and death rates rise, but the two rates never attain equilibrium and thus the cattle population never reaches the limits of its growth. The maximum stocking densities determined by potential ecological carrying capacity are not attained because of the random intervention of exceptionally stressful years. At these times cattle die in unusual numbers and do so at rates which cannot be predicted on the basis of stocking density. In the long run, therefore, non-equilibrium factors tend to be the major influence on cattle population numbers, resulting in populations below potential 'equilibrium' density. However, equilibrium processes are significant during intervening years when cattle populations are high and may be important in the regulation of the cattle population. Over a long time perspective, the semi-arid conditions of southern Zimbabwe apparently result in both non-equilibrium and equilibrium conditions at different times.

As in both Turkana and Borana, the maintenance of livestock in Zimbabwe is contingent upon their mobility and their capacity to exploit variations in the environment. Although not normally characterised as a migratory system of production, cattle in Zimbabwe's communal areas routinely exploit the 'patchy' nature of local vegetation which changes in response to soil differences along drainage systems. In addition to regular, seasonal movements, herds may also engage in long-distance migration out of their home areas in years of exceptional stress.

The common pattern which emerges from the Kenyan, Ethiopian and Zimbabwean case studies is heterogeneity – spatial and temporal variability and its exploitation by pastoral herds and their owners. Discussion thus far has focused on the importance of variability for our understanding of the concept of carrying capacity. We now turn to an examination of the practical problems which this variability poses for the measurement of carrying capacity in the field.

Short-term livestock feed supply and demand

In many instances, attempts to determine carrying capacity are essentially attempts to estimate the levels of livestock output which could be expected from different production systems at different stocking densities. These 'carrying capacity' calculations may be more precisely labelled 'calculations of short-term livestock feed supply and demand', since the focus of analytical interest is not on long term degradation but on the capacity of the system to meet immediate production goals at alternative stocking densities.

In practice, these calculations involve estimating the total edible vegetation produced annually from a specified area and comparing this estimate to the forage consumption requirements of the resident livestock. Two alternative methods for assessing feed supply-demand levels are routinely employed (and a comprehensive review of these methods is provided by de Leeuw and Tothill in this book).

The simplest of these methods is based on estimates of the total edible plant biomass which is produced annually in a rangeland area, routinely expressed in tonnes of dry matter per hectare. Total production is then adjusted by a 'proper use' factor – which routinely varies from about 30% to 45% – representing that proportion of the vegetation which is available for consumption and which the analysts presume can safely be consumed without causing rangeland deterioration in subsequent years. (The issue of how range deterioration is defined and measured is addressed in the next section.) The adjusted production figure is then divided by the feed requirements of an individual animal, and the result expressed in terms of the number of animals which can be sustained per unit of rangeland, an approach employed by de Leeuw et al. in this book.

Calculations of livestock sustainability based on tonnes of dry matter produced per hectare ignore the variable quality of forage as animal feed, a shortcoming which can be redressed by assessing vegetative production in terms of fodder quality rather than quantity. This elaboration of the more standard methods of calculation is applied to Sahelian rangeland productivity in the chapter by de Ridder and Bremen in this book. This chapter forcefully emphasises the depth of research and the understanding of underlying biological processes that must underpin these apparently straightforward attempts at estimation.

Both the precision and utility of evaluating feed supply-demand are, however, open to doubt. With respect to the precision with which estimates can be derived, there is opportunity for significant error at almost every step in the calculation:

(i) The 'proper use factor' is little more than an educated guess, since little is known about the carryover effects of grazing between years; estimated carrying capacities are, moreover, extremely sensitive to alterations in these estimated rates of use. As Bartels et al. note (chapter in this book), the decision to apply a use factor of 45% rather than 30% can increase estimated carrying capacity by half.

(ii) Rainfall-based estimates of biomass production rarely take into account landscape heterogeneity and variability in productivity. For instance, the regression estimator developed by Le Houérou and Hoste (1977) for the Sahel failed to include data points representing the low lying 'bas fonds' areas, where high grass production is found.

(iii) Carrying capacity assessments assume fixed boundaries, but mobility of stock means that these assessments are artificial; on the other hand, it is in practice very difficult to assess 'carrying capacity' in systems where spatially disparate resources are used at different stages of a flexible transhumant cycle.

(iv) Estimation of the amount and kind of forage needed by an animal is not straightforward, especially when several herd species with different feeding habits use the same rangelands, when herd owners pursue different economic objectives, or when livestock feed requirements are derived from research station animals which may not be physiologically or genetically adapted to nutritional stress (Payne, 1965; Western and Finch, 1986).

(v) Compensatory regrowth of grazed and browsed plants, resulting in higher quality and, occasionally, in higher production, is frequently ignored.

These difficulties have occasionally led to estimates which are so obviously wrong as to be embarrassing. Bartels et al., in this book, cite the example of carrying capacity estimates from Somalia which estimate that certain rangelands are chronically overstocked at rates eight times in excess of their capacity, a situation which is biologically impossible.

Conceptual ambiguity, argue Bartels et al. in this book, is compounded by measurement error to the point where carrying capacity estimates do not serve as a reliable tool for planning purposes:

> We have concluded that carrying capacity, as conceived in Western range management, is of questionable validity in livestock production systems in Africa, that it is virtually impossible to estimate it accurately, and that the concept cannot be meaningfully applied in pastoral systems.... The enormous expense devoted to estimating carrying capacity in Sub-Saharan Africa has contributed little to livestock development and has diverted resources from other priorities. Let us admit the problems with the carrying capacity concept, and stop trying to apply it.

And finally, there is the issue of how feed supply-demand estimates might be used. The chapters by de Leeuw et al. and de Ridder and Bremen, in this book, clearly demonstrate the analytical importance of these estimates in attempts to understand the functioning of Sahelian grazing systems. These estimates also colour our perception of African pastoralism, its current condition and development potential. Recent advances in field techniques may diminish the degree of error in these calculations and enhance their accuracy (see especially Bremen and de Ridder, 1991). But analysis is one thing and enforcement is another. As Bartels et al. note in their chapter:

> Though there have been numerous attempts, we know of no case in which a government agency in Africa has successfully persuaded pastoral households, or a pastoral group to voluntarily reduce livestock numbers on rangeland to satisfy an estimated carrying capacity.

Until administrators devise some mechanism for implementing recommended stocking densities, these estimates may provide the background for administrative decision making, but they do not constitute realistic management objectives.

The definition and measurement of rangeland degradation

Range degradation, like the more popular but allied term 'desertification', has been defined in a multitude of contradictory ways (as discussed in ODI, 1977; Sandford, 1983a; Warren and Agnew, 1988). Clear definition is important, both because the issue of rangeland degradation is emotionally charged, and because the meaning which is ascribed to the term largely determines the choice of the diagnostic criteria which are used to measure its occurrence.

In this chapter we equate rangeland degradation with the long-lasting or permanent loss of an economic good, in this case an irreversible decline in livestock production. A formal definition of rangeland degradation consistent with this point of view has been provided by Abel and Blaikie (1989:113), as follows:

> Range degradation is an effectively permanent decline in the rate at which land yields livestock products under a given system of management. 'Effectively' means that natural processes will not rehabilitate the land within a timescale relevant to humans, and that capital or labour invested in rehabilitation are not justified... . This definition excludes *reversible* vegetation changes even if these lead to temporary declines in secondary productivity. It includes effectively irreversible changes in both soils and vegetation.

The phrase 'under a given system of management' merits some elaboration. Different land use systems utilise different components of the natural environment and must maintain those components if they are to be sustainable. To take an obvious example, conservationists will be concerned to maintain the diversity of species present in an area, while commercial wildlife operators may require not species diversity but a plentiful supply of the large game animals upon which they are financially dependent. Degradation assessment, as defined here, does not attempt to determine which of these land use systems is 'best'. It does attempt to assess the capacity of a given management system to maintain those features of the natural environment which are essential for its continued wellbeing.

Potential biological and physical indicators of range degradation have been proposed, including changes in soil, vegetation and livestock condition and output (see Table 1.3). What must now be examined is the extent to which these indicators can identify permanent losses in livestock output which are of genuine concern to pastoral producers. Conventional range management has relied on vegetation indicators to assess range degradation. Whether these indicators are also reliable measures of permanent declines in economic output from Africa's rangelands is, however, open to doubt.

The initial problem is to decide what we want to measure: declining productivity or vegetation change. Given the essentially economic definition of degradation employed here, vegetation change is of no intrinsic interest unless it also provides reliable evidence of changes in livestock productivity. The high stocking rates which are maintained by some pastoralists will, almost certainly, alter 'pristine' or 'climax' vegetation, in equilibrium grazing systems or in areas of stock concentration around water points or settlements (Coppock in this book; de Leeuw et al. in this book; Grouzis, 1990). These ranges will tend to be in poor condition, if range condition is successionally defined, but, as Wilson and Tupper have observed, 'agriculture in general is based on the modification or replacement of natural vegetation, and rangeland, although only partially modified, must be assessed on the same basis' (1982:689). Very few agriculturalists would conclude that an English sheep paddock or a Javanese rice paddy were 'degraded' solely because, several centuries previously, they had replaced a temperate or a tropical forest. A more important question is whether any of these agricultural systems, including pastoral systems of range exploitation and the man-made environments they have created, are sustainable in the long run.

Direct examination of rangeland vegetation does not provide a simple answer to this question. Large fluctuations in species composition, plant biomass and cover are characteristic of arid and semi-arid rangelands subjected to erratic rainfall. Because the vegetation in these areas is continuously disturbed, it has adapted to disturbance and possesses an enhanced capacity to recover from disturbance (Walker et al., 1981). The productivity and composition of such rangelands may be unstable in the short run, but resilient over the long term (Holling, 1973).

In such an environment, degradation could be said to occur only when the vegetation had crossed, or was at risk of crossing, critical thresholds which prevent or severely inhibit its subsequent return to a more productive state. In practice, the problem is to distinguish between drought induced fluctuations and permanent changes in vegetation states (Grouzis, 1990). Current knowledge of the dynamics of savanna ecosystems frequently does not permit this distinction to be made with confidence, although future research may eventually clarify the issue (Friedel, 1991; Laycock, 1991). As a result, it has, thus far, proved very difficult to differentiate between permanent human-induced 'degradation', as opposed to temporary rainfall-induced vegetation change (Alchrona, 1989; Warren and Agnew, 1989; Tucker et al., 1991).

Table 1.3: Biophysical indicators of degradation

Soil changes

- Decreased fertility
- Decreased water holding capacity
- Decreased infiltration
- Soil loss significantly in excess of soil formation

Vegetation changes

- Changes in vegetation productivity over time, unrelated to rainfall patterns
- Changes in vegetation cover
- Changes of plant species composition of use to animals
- Shifts between vegetation transition states that result in decreased fodder (eg. severe bush encroachment)

Livestock production

- Condition scoring of animals
- Calving rates and death rates (population models)
- Milk yields

Source: Woburn Rangeland Workshop discussions

The need for caution in presuming degradation and care in measuring it are well illustrated by Tapson's chapter in this book, in his assessment of the extent of degradation in Kwazulu, South Africa. His analysis relies on a combination of administrative records of unusual historical depth and experimental research of exceptional scientific rigour and coverage, relative to the information available about most of Africa's rangelands. Tapson is able to locate at least some data on all the basic indices of biophysical degradation cited in Table 1.3 – changes in soil, vegetation and livestock production. Despite this relatively robust data base, his conclusions exhibit a modesty which is in marked contrast to the unqualified and apocalyptic generalisations sometimes made regarding the link between overstocking and rangeland degradation:

> Even under controlled research conditions, the assumed relationships upon which the present understanding of the dynamics of grasslands is based, are not sufficiently consistent to be used as the basis for policy prescriptions. This applies in the case of a grassland resource such as that employed by Zulu cattle owners.... It is argued here that in fact the technical evidence is so fragile it does not of itself present a valid case for destocking.

Tapson's conclusions are broadly endorsed in analyses by Biot (chapter in this book) and Abel (chapter in this book) of rangeland degradation in the neighbouring southern African country of Botswana. Given the problems of using vegetation change as an indicator of irreversible rangeland degradation, these chapters explore the possibility of assessing degradation in terms of soil loss and other deleterious changes in soil chemistry and physical properties. The challenge, in this case, is to develop techniques for measuring and modelling soil loss, the focus of Biot's contribution, and to relate these measures to economically significant changes in livestock output, the objective of Abel's analysis.

Biot, in his chapter, presents a soil loss model for a portion of the hardveld rangelands of eastern Botswana. In this eroding landscape, as in much of arid and semi-arid Africa, rates of soil loss are greater than rates of soil formation, even with zero use. While human use might accelerate ongoing processes, stopping environmental change is not an option. Biot uses the concept of 'soil life' or 'residual soil suitability' to express the length of time a given level of output from the land can be maintained under different intensities of grazing. His estimation techniques provide an unexpectedly optimistic picture of soil loss on Botswana's communal rangelands. At the stocking densities prevailing at the time of his study, he estimates the residual soil life in his study area to be over 400 years. Environmental change is certainly taking place in Botswana, but not at the catastrophic rates routinely depicted (Cook, 1983).

Biot's results cannot be generalised; they pertain to only one landscape and one management system. What may be generalised are his modelling techniques. He explores the potential of these techniques by comparing rates of soil loss for hypothetical rangeland systems in the semi-arid, wet and dry tropics. As might be expected, this comparison demonstrates that landscapes respond very differently to grazing pressure depending on factors such as rainfall, slope, soil texture, and vegetative cover. While Biot's conclusions are at this stage only indicative, they suggest that additional field work may make it possible to quantify both the risk and rate of soil loss from rangelands under different environmental and management conditions.

Abel (chapter in this book) builds on Biot's analysis of erosion in eastern Botswana in an attempt to further specify what might be an 'economically acceptable' rate of degradation. Abel compares the economic costs to Botswana herd owners of maintaining current levels of soil loss versus reducing those levels. He bases his comparison on a model which predicts the immediate and long-term effects of two different stocking rates, the current stocking rate in the communal areas of eastern Botswana versus the lower, government-recommended stocking rate for these areas.

Based on earlier estimations of herd productivity at these two stocking densities, Abel concludes that the lower, recommended density would significantly reduce the aggregate productivity of the communal herd and do so at considerable collective cost to herd owners. He also shows that the current (high) and recommended (low) stocking densities produced virtually identical levels of soil loss between 1978 and 1988, given the pattern of rainfall in that period. Put simply, the immediate costs to producers of destocking would be heavy, while the long-term gains in reduced range degradation would be slight. In eastern Botswana, destocking is not worth it.

In a topographically complex landscape, soil lost from eroding areas, such as slopes, may be transported and subsequently redeposited elsewhere within the landscape, resulting in a relocation rather than an absolute decline in soil resources, plant growth and grazing activity. Abel and Biot's models are restricted to an estimation of slope erosion. Net soil loss from the hardveld landscape in eastern Botswana is a fraction, possibly only 20% to 25%, of the slope erosion estimated by Biot (Abel and Stocking, 1987; Biot, in this book).

Stafford Smith and Pickup (chapter in this book) present techniques for the analysis of such processes of soil and productivity relocation. Their material is drawn from ranching areas of arid Australia, areas which experience climatic fluctuations similar to arid African and where ranches are large enough for livestock movements to replicate some of the patterns characteristic of Africa's open rangelands.

Stafford Smith and Pickup begin with a comprehensive review of alternative conceptual models and mechanistic or simulation models of vegetation change. They argue that all these models - including the older Clementsian and the newer state-and-transition models - have difficulty dealing with the simple observation that soil is not only eroded, but also transported and deposited. Soil loss at one site generally means that soil is accumulated at another, such that primary productivity may not only be lost or gained, but also relocated within a landscape. Models of vegetation change may provide little insight if they are insensitive to the flows over time of nutrients, water and soil between sample points on the landscape, or deal with mean values averaged across a landscape rather than with the changing spatial patterns of variance within a landscape.

Based on their previous work, Stafford Smith and Pickup provide techniques for incorporating spatial variables into rangeland and vegetation assessment. Given the mobility of both human and animal populations in Africa, and the capacity of mobile populations to exploit spatial heterogeneity, these techniques would seem to offer an improved methodology for understanding both environmental change and the response of African pastoralists to change.

A classification of rangeland types: implications for management

The distinction between equilibrium and non-equilibrium grazing systems calls for a rethinking of rangeland classification. In practice, range managers need to be able to distinguish between those types of rangeland in which non-equilibrium models are appropriate and those in which conventional successional interpretations, and concepts like carrying capacity, are still relevant.

Many different classifications have been used to distinguish African savanna types. Grassland ecologists have differentiated savannas according to species composition (Acocks, 1953; Rattray, 1957; Pratt and Gwynne, 1977); others have classified savannas in relation to topographical variations in the landscape (Milne, 1947; Morison et al., 1948). Only recently have more analytical classifications, based on models of savanna functioning, emerged (Frost et al., 1986; Solbrig, 1991). These models ascribe overriding importance to soil fertility and moisture in the genesis of different forms of savanna vegetation.

In general, primary production and animal density in a savanna are positively correlated with mean annual rainfall (Coe et al., 1976; Le Houérou and Hoste, 1977; Rutherford, 1978; Deshmukh, 1984). The simple relationship between high animal density, high levels of primary production and high rainfall is complicated, however, by a third variable – soil type as influenced by base geology. Bell has provided empirical evidence that, at comparable rainfall levels, savannas with nutrient-rich or poor soils support different types of vegetation, and variable densities and kinds of herbivores (Bell, 1982, 1984).

Implicit in Bell's analysis is a functional classification of savanna types based on various permutations of available soil moisture and soil nutrients (Frost et al., 1986). This classification is presented in Fig. 1.5.

Soil nutrient availability, the horizontal axis in Fig.1.5, is influenced by parent geology, and by nutrient transport from weathering and water movement. The availability of moisture for plant growth, the vertical axis, is determined by total rainfall levels and distribution, soil physical properties (particularly infiltration rates) and topography. Various combinations in plant available moisture and nutrients create the major vegetation types noted in Fig. 1.5.

The following section discusses the implications of this savanna classification for the management of African pastoral areas. We ask what the classification system tells us about the likelihood of equilibrium or non-equilibrium dynamics, expected patterns of degradation for different rangeland types and the implications for feed resource management.

Figure 1.5: Hypothetical distribution of savanna types in relation to the main determinants of savannas

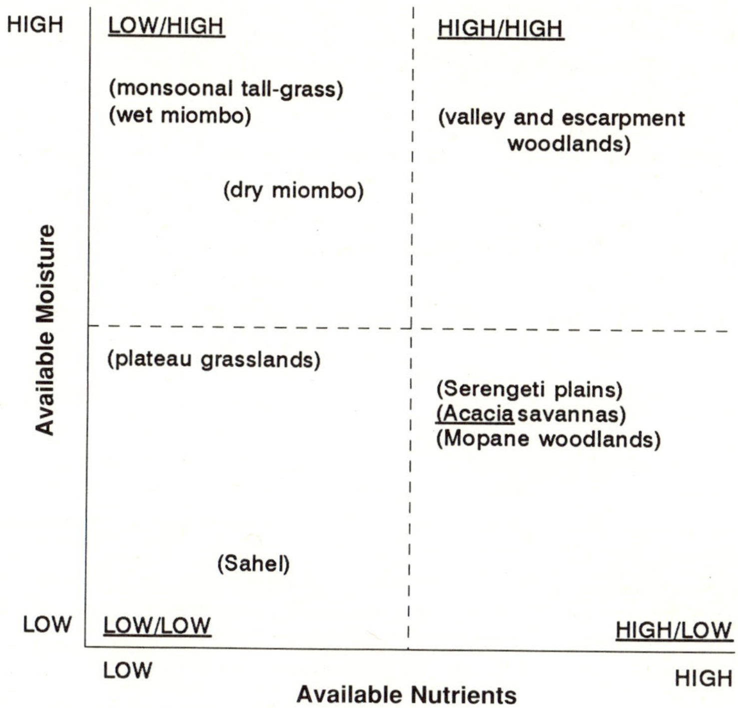

Source: Reproduced with modification from Frost et al. (1986)

Degradation in equilibrium and non-equilibrium environments

As rainfall becomes low and erratic (vertical axis on Fig. 1.5), both primary productivity and livestock populations will fluctuate widely and non-equilibrium dynamics will predominate (Ellis and Swift, 1988). Conversely, relatively wet savanna areas with stable rainfall regimes may be able to sustain livestock densities which have a significant impact on plant biomass and species composition, the classic equilibrium situation (Coppock, chapter in this book). In areas where both wet and dry periods occur, there may be a shift between equilibrium and non-equilibrium dynamics over time (Scoones, chapter in this book).

The instability inherent in certain climatic regimes may, however, be exacerbated, or dampened, according to soil type (horizontal axis on Fig. 1.5). On fertile clay soils, levels of primary production are closely correlated with, and as variable as, annual rainfall levels. This instability results from a combination of adequate soil fertility, which induces high levels of plant growth when water is sufficient, combined with the poor water infiltration and retention capacity of clay, which severely limits plant growth when water is insufficient. Coarse but nutrient deficient soils show the opposite pattern – relatively stable plant growth constrained, during periods of good rainfall, by the availability of nutrients, but maintained, at low rainfall levels, by the capacity of the soil to admit and hold water (Dye and Spear, 1982).

Soil physical and chemical properties may also influence the way in which different range types respond to grazing pressure. Grazing pressure on heavy textured soils has a significant effect on infiltration through soil capping, compacting of soil structure, removing of litter, and decreasing the density of perennial grass tufts (eg. Kelly and Walker, 1976; O'Connor, 1985). Under heavy grazing pressure, increased run-off and decreased infiltration can result in undesirable changes in vegetation states leading to the creation of poor quality open grassland or encroached woodland (Walker et al., 1981; Grouzis, 1990). By contrast, sandy nutrient-poor savanna soils, and the vegetation they support, appear to be more resilient to herbivore impact (see Barnes, 1965 for Zimbabwe). As a result of higher infiltration in sandy soil, the grass layer tends to be insufficient to out-compete the woody component, and, with the exception of extremely low rainfall areas, a woody-grass vegetation is relatively stable.

Finally, the positive correlation between soil fertility and plant palatability may also influence the stability of the grazing system. Except in very low rainfall areas (as in the northern Sahel), poor soils support a vegetation characterised by woodland and grassland of low nutritional value from grazing animals. Relatively low densities of herbivores are able to survive in this environment and their grazing may have only a marginal impact on plant biomass and the relative balance of woody and herbaceous species. By contrast, savannas with higher quality soils support a higher density (and a greater diversity of wild herbivore species) because of the better quality feed resource. Under these conditions, stocking densities may be sufficiently high to suppress the standing crop of herbaceous material and/or suppress woodland and encourage grassland (Bell, 1982, 1984).

In sum, climatic instability, manifested in low annual rainfall levels and high coefficients of rainfall variation (Caughley et al., 1987), is the probably the most reliable single indicator of the shift from equilibrium to non-equilibrium grazing systems. Soil factors may nonetheless suppress or exaggerate the effects of an erratic climate. Sandy, nutrient-poor soils produce vegetation which is relatively stable in its productivity, unpalatable, and resistant to herbivore grazing pressure. Range types on these soils may be relatively less exposed to degradation, when low and erratic rainfall suppresses livestock numbers (the low/low quadrant in Fig. 1.5) or when high rainfall levels produce unpalatable vegetation and low stock densities relative to biomass production (the low/high quadrant in Fig. 1.5). Savanna types on fertile clay soils exhibit the opposite characteristics: instability in biomass production (under fluctuating rainfall), high feed palatability and high but potentially variable stock densities. Because the soils are prone to compaction, both soils and associated vegetation may be susceptible to degradation if rainfall is reliable enough to sustain high stock densities (the high/high quadrant in Fig. 1.5).

Feed resource management

Fodder palatability (quality) and biomass (quantity) both vary with changes in available plant moisture and nutrients. With respect to grasses, there is generally an inverse relationship between biomass production and palatability. Palatability tends to increase with improved soil fertility and/or reduced soil moisture. Under very dry conditions, annuals are dominant. The production of fodder biomass shows an opposite trend, with higher biomass production found in wetter rangeland areas where perennial grasses dominate.

The relative balance of trees and grasses also depends on soil properties, and the abundance of water. In heavy soils, the upper soil layer may retain a significant proportion of incoming water, allowing the growth of a vigourous grass layer which can inhibit the regeneration of trees. At similar rainfall levels but on lighter soils, most water penetrates to the sub-soil and the vegetation may be dominated by trees, which have roots which are deep enough to utilise this source of water (Walter, 1971; Walker et al., 1981; Knoop and Walker, 1985; Coppock, chapter in this book).

Feed resource management will vary according to fodder quality and quantity in different range types. For instance, the use of the tree layer as a fodder resource will reflect the availability and quality of alternative feed sources. In nutrient poor grasslands, browse resources may be important in supplying high quality feed to livestock at particular times of the year. Alternatively, in nutrient rich arid areas, grass biomass production may be highly variable and feed quantity may be an important seasonal constraint. In these situations browse may provide bulk feed when grass biomass is insufficient.

The nature of the fodder resource also affects the way animals are managed within and moved between different range types. African rangelands are ecologically heterogenous at a variety of different spatial scales. Local variability is important because it occurs over distances which livestock can walk. It would appear that animal movements – seasonal, annual and daily, local and long-distance, by both wild and domestic herbivores – systematically exploit the environmental discontinuities summarised in Fig. 1.5 (Scoones, 1989a; Breman and de Wit, 1983; McNaughton, 1985; McNaughton and Georgiadis, 1986). Different parts of the landscape may be critical in offsetting particular constraints. In the Sahel, for example, livestock are moved from low quality, high biomass range types in the dry season, to high quality, low biomass range types in the wet season (Breman and de Wit, 1982). In Turkana a heterogeneous rangeland resource is partitioned among a number of different domestic herd species, which follow distinctive seasonal patterns of movement and resource utilisation (Coppock et al., 1986). In semi-arid Zimbabwe, movement to relatively small but critical areas of high production, along rivers, streams or drainage lines, can be critical in sustaining livestock populations in the dry season, while top lands are grazed following the rains (Scoones, 1989a).

In all these cases, a vital step in understanding and possibly improving rangeland management strategies is the identification of key resources areas which redress critical constraints for livestock production for a particular range type. Analysis of constraints according to the interactions outlined in Fig. 1.5. will assist in identifying key resources for different range types.

Opportunistic management

International development agencies and African governments have devoted considerable effort to the suppression of pastoral techniques of land and livestock management. These programmes were undertaken on the presumption that pastoralism was inherently unproductive and ecologically destructive and, hence, required radical reform. Current empirical research supports none of these presumptions.

With respect to herd productivity, comparative studies of ranch and pastoral herd output in West Africa (Breman and de Wit, 1983), Southern Africa (de Ridder and Wagenaar, 1986; Abel, in this book) and East Africa (Cossins, 1985; Western, 1982) demonstrate that pastoralism either equals or exceeds the productivity per unit land area of commercial ranching in comparable ecological environments. Any attempt to improve on the productivity of African pastoralism can, at best, aim to marginally increase already high levels of output.

The work reviewed here makes much the same point with respect to pastoral methods of range management. This chapter documents a convergence between pastoral techniques of range exploitation and recent developments in scientific range ecology. This convergence does not constitute a blanket endorsement of the positive ecological impact of African pastoralism. It is now clear, however, that pastoral land use practices are an effective response to the exigencies of a difficult natural environment, and that the development of livestock production in dry Africa requires the refinement and adjustment of these practices to changing circumstances, not their outright elimination.

Not confined to an arbitrarily demarcated ranch and with limited access to industrial inputs, African pastoralists have had little capacity or imperative to control localised fluctuations in rangeland productivity. They have, instead, adapted to instability. This attempt to exploit environmental instability and contingent events may be characterised as 'opportunistic management' (Sandford, 1983a; Westoby et al., 1989). High but fluctuating stocking rates and migratory patterns of forage exploitation are recurrent features of pastoral opportunism. Any systematic attempt to build upon pastoral husbandry practices and incorporate them into formal development programmes must examine the utility, and the limitations, of these management techniques.

With respect to specific management and policy issues in particular local settings, the contributions to this book offer several suggestions. The discussion of rangeland classification in this overview chapter has specified the kinds of natural environments which are suited to conventional or opportunistic management approaches. A revised assessment of the merits of opportunism will, however, affect almost all aspects of pastoral development policy in dry Africa. In this closing section, we briefly explore some of the wider implications of opportunistic rangeland management for the redesign of these policies.

Sandford's analysis of the relative advantages of conservative or opportunistic stocking strategies provides a useful point of departure. Briefly, Sandford distinguishes between a conservative stocking strategy in which a 'constant number of livestock graze an area through good and bad years alike' versus an opportunistic strategy 'in which the number of livestock grazing is continuously adjusted according to the current availability of forage' (1983a:38).

Because the intention is to hold animal numbers constant, conservative stocking rates are determined by the number of animals which can be maintained during periods of low forage availability. Conservatism is a matter of degree, but a conservative stocking rate always carries a cost – the forage which cannot be consumed and the livestock production which is thereby foregone in good years because livestock numbers are insufficient to consume all available feed. As Sandford has shown, this cost increases as the variability of rainfall increases and to the extent that managers adopt safer, more conservative stocking rates.

Opportunistic or variable stocking rates reduce the problem of unconsumed, and thereby surplus, forage in good years, but present potential problems of surplus stock in poor years. Livestock development programmes based on opportunism would not attempt to suppress these fluctuations in livestock numbers, but to exploit them by developing mechanisms to promptly and profitably remove stock when it does not rain, what Sandford (1983a) has characterised as *efficient* opportunism. In this framework, livestock development policy would not be judged by its success in preventing periodic crashes in livestock numbers, which are inevitable, but by the appropriateness of its response to these crashes. At least three aspects of pastoral development policy would require revision in light of this changed objective.

Livestock marketing

Livestock sales are one obvious means to achieve rapid destocking, and livestock marketing would play an important role in an opportunistic policy towards rangeland management, as it has done in conventional livestock development programmes. However, the futile attempt to maintain constant levels of stock sales in order to prevent herd growth would be de-emphasised, and attention would shift instead to the design of marketing systems which can accommodate massive and unpredictable shifts in levels of throughput. A detailed examination of how this kind of marketing system might operate lies well beyond the scope of the present discussion, but it is clear that the organisation, infrastructural requirements, performance criteria and financing of these systems would depart considerably from past attempts to improve livestock marketing.

Herd movement and land tenure

Livestock movement is a second means to adjust local imbalances in stock numbers and forage availability. Opportunistic management would seek to maintain mobility as a production strategy and to adapt this characteristic feature of pastoral nomadism to changing economic and institutional conditions. A new approach to pastoral land tenure would need to be a critical component of this effort.

Previous attempts to reform pastoral tenure rights have concentrated on delimiting bounded areas and restricting livestock to those areas. Since it was assumed that pastoralists would eventually settle on something like a ranch, little official effort was devoted to the question of maintaining pastoral tenure rights to key land resources which were intermittently used and not continuously occupied. To the extent that they are based on the use of force, customary pastoral techniques for maintaining these rights are incompatible with civil administration. The result has been the widespread deterioration of pastoral rights to scattered but highly productive categories of rangeland throughout dry Africa.

Any official attempt to foster opportunism by maintaining livestock mobility would require the development of legal formats capable of providing security of tenure while permitting flexibility of use patterns. This will be no easy task. Models for this kind of tenure system are not readily available from pastoral areas of industrialised countries, which have themselves had a chequered record with respect to the promulgation of appropriate pastoral tenure legislation.

Pastoral administration

Finally, there is the question of who manages an opportunistic management system. Conventional range management in dry Africa has been highly interventionist. It has generated much bureaucracy, but little effective action. The non-equilibrium view of range ecology suggests an alternative management model which relies on limited but focused interventions coinciding with key events, interspersed with long periods of minimal administrative interference. This suggests less rather than more centralised regulation, the devolution of control over local resources to producers and producer groups, and a shift in emphasis from enforcement to monitoring critical developments and servicing local needs (Swift, 1990).

By definition, there can be no set blueprint for opportunism. Any attempt to systematically develop it would require a development programme tailored to particular settings. Pastoral communities are uniquely qualified to undertake these local adjustments and refinements; scientific recognition of the competence of these communities as land managers is a first step in this direction.

2

Climate Variability, Ecosystem Stability, and the Implications for Range and Livestock Development*

James E. Ellis, Michael B. Coughenour, and David M. Swift

Some long-term trends in ecological theory

In mid-1990, the *New York Times* carried an article entitled 'Balance of Nature Concept Challenged'. This article reported on a symposium at the annual meeting of the Ecological Society of America (ESA), where evidence was presented that 'change and turmoil, rather than constancy and balance are the rule' in nature. This was truly newsworthy, because, for almost 200 years, the concept of the equilibrial nature of natural systems, i.e., the balance of nature, has dominated ecological and evolutionary theory and thought as well as natural resource management. In 1798, Thomas Malthus, 'the first professional economist', initiated what has turned out to be one of the most durable of ecological theorems, namely, that populations are regulated by their food supply, through density-dependent feed-back relationships. Regarding evolutionary theory, Darwin and Wallace simultaneously presented, in the late 1880s, their view of the gradual evolution of species through the orderly processes of competition and natural selection (Darwin, 1859).

Both these evolutionary and ecological precepts are based on an assumption of equilibrium. That is, that conditions outside the system of interest are relatively stable over time, allowing the internal processes of the system to play out or equilibrate, and to regulate system structure and dynamics. Thus, the 1990 ESA symposium, in questioning this 'balance of nature', challenged one of the oldest and most fundamental premises in ecology and evolutionary biology. In fact, challenges to the equilibrial concept and its derivatives have been part and parcel of scientific debates for a long time and many of the dissenting views originated from the field of resource management. And in the last 10-20 years these dissenting voices have coalesced into an alternative concept.

Every range, forest, fisheries, and wildlife ecologist can appreciate that equilibrial concepts have (in addition to their impact on ecological theory) greatly influenced the management of natural resources. Range and livestock management has, for 65 years, been built around the concept of range condition class and the practices of determining carrying capacities and manipulating livestock numbers and grazing seasons to influence range condition. This managerial approach is, of course, derived from the equilibrial or climax concept of Clementsian succession (Clements, 1916).

---

* This paper is also published in Cincotta et al. 1991.

Although succession, as Clements presented it, was the target of criticisms over many years (Gleason, 1927; Whittaker, 1951; Drury and Nisbet, 1973), it was still the focal concept of ecosystem theory according to Margalef (1968) in the late 1960s, and succession provided the central theme of Eugene Odum's synthetic treatise on the 'strategy of ecosystem development' published in 1969. In this paper, Odum suggested that succession theory might be the 'basis for resolving man's present environmental crisis'. This clearly provided a vote of confidence from a distinguished ecologist for the equilibrial nature of ecosystems and the appropriateness of succession as a foundation for natural resource management. Odum's paper represents a watershed event in ecology, not only because it synthesised half a century of previous successional and ecosystem theory, but also because it marks the beginning of the period (1970-80) when dissatisfaction with this balance-of-nature concept began to condense into an alternative (non-equilibrial) view of the dynamics of ecosystems and the evolution of species.

In the early 1970s, papers began to appear which emphasised neither the systematic and equilibrial nature of ecosystem interactions nor the predictable patterns of successional and evolutionary change. Rather, a different view of nature began to emerge which stressed the random occurrence of external events and the idiosyncratic and unpredictable effects of externalities on ecosystem dynamics and on populations and species. One of the most provocative ideas in evolutionary theory since Darwin was presented by Eldredge and Gould in 1972. In their paper on punctuated equilibria, they hypothesised that extraneous, large-scale, random 'events' have had more influence on the course of evolutionary history than have competitive interactions and the slow, equilibrial fine-tuning of Darwinian natural selection.

Ecologists were also reconsidering the equilibrial nature of ecological systems at this time. In 1973, C.S. Holling reviewed the concepts of stability and resilience in ecological systems; he advocated shifting attention from equilibrial states and interactions to exploring the conditions necessary for system persistence. In Holling's view, resilience to external perturbation and persistence in a changing world are the crucial characteristics of ecological systems. 'Natural, undisturbed systems are likely to be continually in a transient state; they will be equally so under the influence of man' (Holling, 1973:2). Also in 1973, Noy-Meir postulated that dry ecosystems are controlled more by climatic events and sequences that by equilibrating interactions among the biotic components of the system, a distinctly non-equilibrial assertion. However, in later papers he emphasised the role of biotic interactions in causing systems to shift from one stable state to another (Noy-Meir, 1973, 1975; Walker and Noy-Meir, 1982).

Evidence from another set of arid and semi-arid ecosystems caused Weins (1977) to propose that climate, rather than competitive interactions, regulated avian community structure in the grasslands and shrublands of North America. He went on to make the prescient prediction that ecosystems exist along a gradient of conditions, with biotic interactions controlling community structure where abiotic variability is small (classical equilibrial conditions), while these same abiotic effects come to control communities when climate is highly variable (non-equilibrial conditions) (Weins, 1984).

The relevance of these concepts to extensive range, livestock, and wildlife systems has become evident in the 1980s. Focusing on African pastoralists and pastoralism, Lamprey (1983) postulated that African savanna ecosystems are equilibrial in nature, and where native wildlife systems exist, stable, equilibrial density-dependent interactions predominate. This premise is generally well supported by the results of several long-term ecological studies conducted at the Serengeti Research Institute (Sinclair and Norton-Griffiths, 1979; McNaughton, 1985; Dublin et al., 1990). Lamprey further suggested that pastoralists often destabilise and degrade these potentially equilibrial African ecosystems, through overstocking and overgrazing. It is important to note that Lamprey's experience was predominantly in Tanzania, in areas of relatively mesic (500-1000 mm rainfall) grazing systems.

A very different view of African pastoral systems was expressed at the same time by the economist Stephen Sandford (1983a). He made the important observation that as rainfall variability increases, so do the opportunity costs of a conservative grazing strategy. Such a conservative strategy would maintain livestock at a safe density, dictated by the long-term carrying capacity of the system. Sandford argued that an opportunistic strategy, in which stocking rates track plant production, was appropriate under highly unpredictable rainfall conditions. His view of pastoral livestock systems is distinctly non-equilibrial and reflects the dynamic nature of arid African ecosystems. Lamprey's ideas are clearly equilibrial and reflective of his experience in the more mesic grazing lands of Tanzania.

An extremely thorough analysis of an arid (>250 mm rainfall) grazing system in New South Wales, Australia, was conducted by Caughley and his associates (Caughley et al., 1987). They evaluated the interactions among climate, plants, sheep, and kangaroos and characterized this arid system as highly dynamic over time. They concluded that 'highly variable systems differ in both degree and kind from slightly variable systems' (Shepherd and Caughley, 1987:194) and suggest that where the coefficient of variation (CV) of annual rainfall exceeds about 30%, then the long-term performance of a system is better characterised in terms of its variability than by measures of mean values. Although they eschew the terms equilibrial and non-equilibrial, their results clearly reflect the climate-driven, dynamic sort of ecosystem behaviour that some have characterized as non-equilibrial. They proposed that, in this kind of system, the concept of an ecological carrying capacity is largely an abstraction and not very useful.

Our own work in an arid (300-400 mm precipitation) region of northern Kenya (summarised by Ellis and Swift, 1988) reached similar conclusions. We hypothesised that the dynamics of the Turkana pastoral ecosystem are largely controlled by frequent drought perturbations and that the system operates far from equilibrium most of the time. Under these circumstances, it seems unlikely that pastoralists would routinely overstock and overgraze as suggested by Lamprey (1983) and many others. We also questioned the applicability of traditional range management and livestock development procedures under such non-equilibrial conditions.

Alternatively, recent results from a long-term study of Borana pastoralists, conducted in the Southern Ethiopian rangelands by ILCA scientists, support the idea of equilibrial dynamics in pastoral ecosystems. Coppock's chapter in this book reports that in the relatively mesic (400-700 mm rainfall) Ethiopian highlands droughts are infrequent, livestock appear to be near carrying capacity much of the time, and there seems to be a cyclic pattern of overgrazing, bush encroachment and eventual recovery. Under these circumstances it is clear that 'Carrying capacity, successional trends, and density-dependent population dynamics are viable concepts...' (see page 60).

These varied results, ideas, hypotheses, and theories expounded over the last twenty years lead us to the following conclusions:

(i) There is a long-standing and well-established theoretical basis supporting the idea of equilibrial dynamics in ecosystems.

(ii) Our traditional approaches to range management and to livestock development are founded on this theoretical base.

(iii) For many years ecologists and other scientists, especially those working in arid environments, have questioned this theoretical construct. This dissension has led to the development of an alternative theoretical construct, the non-equilibrial hypothesis, to explain the dynamic behaviour of those dry ecosystems.

(iv) Results from studies conducted in arid ecosystems over the last twenty years seem to offer strong support to the non-equilibrial hypothesis.

(v) Up to now, there has not been a well-orchestrated attempt to build a set of management procedures explicitly based on the non-equilibrial hypotheses.

(vi) There remains a division of opinion regarding these alternative hypotheses and appropriate managerial procedures. It is quite clear that Weins (1984) was right when he suggested that we might discover a whole spectrum of ecosystem dynamic behaviour patterns ranging from equilibrial to non-equilibrial.

We take encouragement from these varied hypotheses, studies, theories, and ideas. They suggest that ecologists do have the capacity to learn, and that new knowledge and ideas will improve our ability to manage and develop rangeland ecosystems. At this time it seems particularly important to do a better job of identifying the long-term patterns of rangeland dynamics for specific types of ecosystems, and then adapting management and development techniques to fit these dynamic patterns.

Climate variability and patterns of arid ecosystem dynamics: The northern Kenya case

Our work in Turkana District, Northern Kenya, provides an example of a region where climate variability is high and ecosystem behaviour is quite dynamic. Our decade-long study, the South Turkana Ecosystem Project (STEP), has provided some useful insights into how this climatic variability drives ecosystem processes, including the flow of energy from plants to livestock to pastoralists (Coughenour et al., 1985). We have also seen how climate perturbations, in the form of recurrent drought, control plant biomass, livestock numbers, and human exploitation strategies (Ellis et al., 1987). Ten years may be a long time to study a single ecosystem; however, it is a brief period when the focus of interest is the long-term behaviour of the entire system. For this reason we attempted to simulate the long-term behaviour of this ecosystem, using actual precipitation records to drive a simulation model developed from our studies (Coughenour et al., 1990; Swift in Ellis et al., 1987).

We tested the plant production portion of this model by driving the model with actual rainfall records for the period 1981-8. The simulated production sequence was then plotted against remote-sensed normalised difference vegetation index (NDVI) values for the study area during the same period, to evaluate the performance of the model (Fig. 2.1). Next, the model was exercised using a 65-year precipitation record from the only station with long-term weather data in Turkana; this is the district headquarters, Lodwar. The mean annual rainfall for Lodwar over this 65-year period is near 180 mm (Fig. 2.2), but the patterns of variation are more interesting than the mean. The CV of the annual means for this sequence is greater than 60%. Fig. 2.3 portrays

Figure 2.1: NDVI vs. modelled biomass

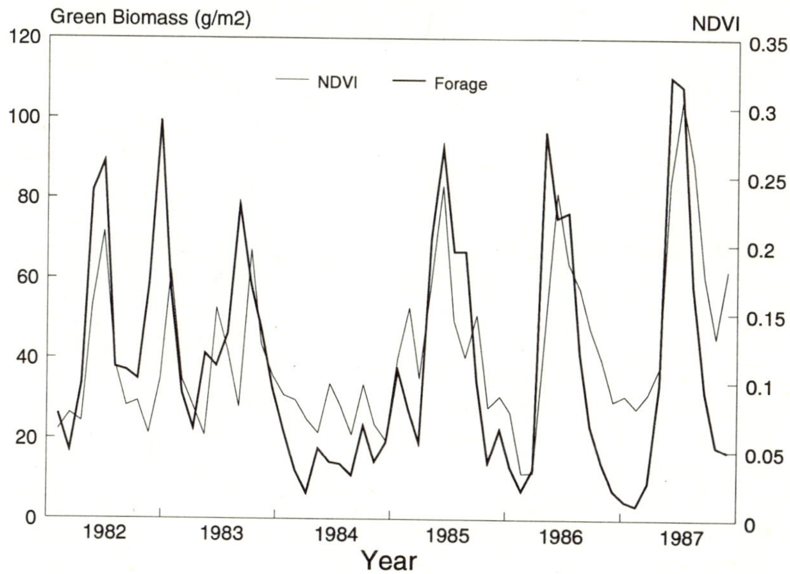

Figure 2.2: Annual rainfall Lodwar, Kenya, 1928-1988, with 3-year running mean

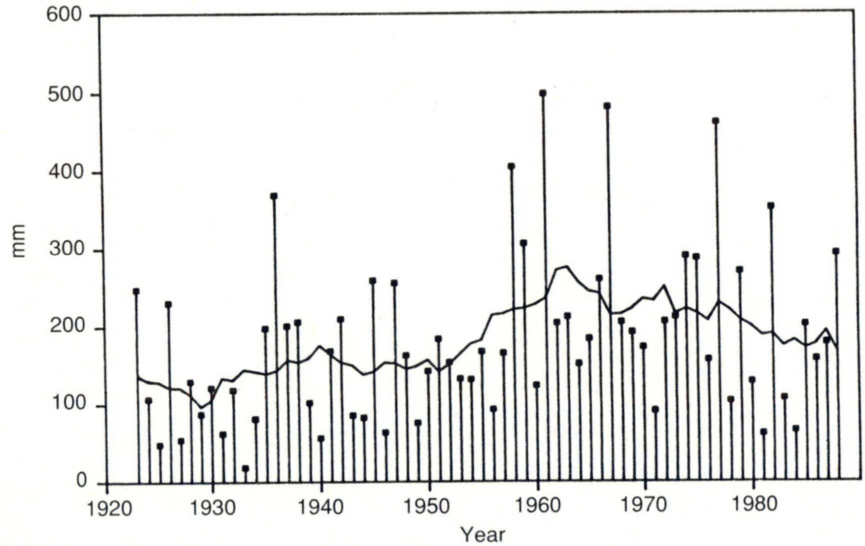

these data as the percentage deviation from the long-term mean. This graph shows a long dry period of roughly 35 years duration, followed by a 20-year sequence of relatively high rainfall with few droughts, and finally a 9-year period with both severe droughts and high rainfall years. Simulation runs covering this period (Fig. 2.4a-d) suggest a very dynamic ecosystem. During the drought of the 1920s and 1930s, the model shows uniformly poor primary production with a resulting very poor livestock condition index. Although modelled plant production improved significantly during the next twenty years, the livestock condition index remained volatile, improving during good rainfall years and declining during droughts. Over the next two decades, until 1979, rainfall was generally good. There was only one drought during the period, and the simulated livestock condition index remained high. It seems possible that livestock numbers may have approached the ecological carrying capacity of the ecosystem at this time. Turkana informants tell us that some pastoralists became semi-sedentary in the early 1970s. This benevolent period ended in 1979-80 at the time when our study began. Since 1980, rainfall patterns have been erratic and the decade has been marked by two droughts, one of which caused livestock mortality on the order of 50% of the regional herd. The modelled livestock condition index reflects the effect of this 2-year drought which occurred early in the decade and resulted in the death of half the regional livestock herd. Interestingly, the more severe but single year 1984 drought caused only a mild decrease in the model livestock condition index; in fact, very few livestock died during the 1984 drought, because of its short duration.

Figure 2.3: Percentage deviation from long-term mean, rainfall at Lodwar, Kenya, 1923-1986

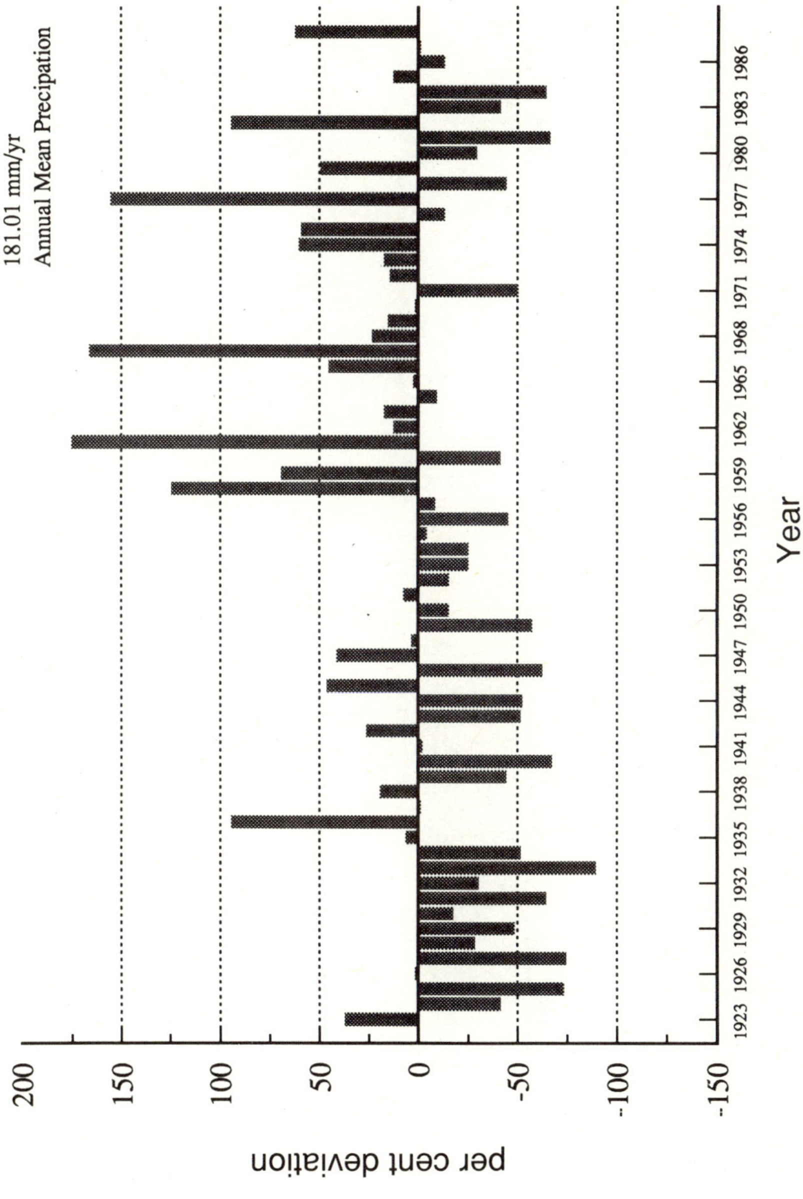

Figure 2.4: Simulated condition index, Lodwar, Kenya, 1926-1988

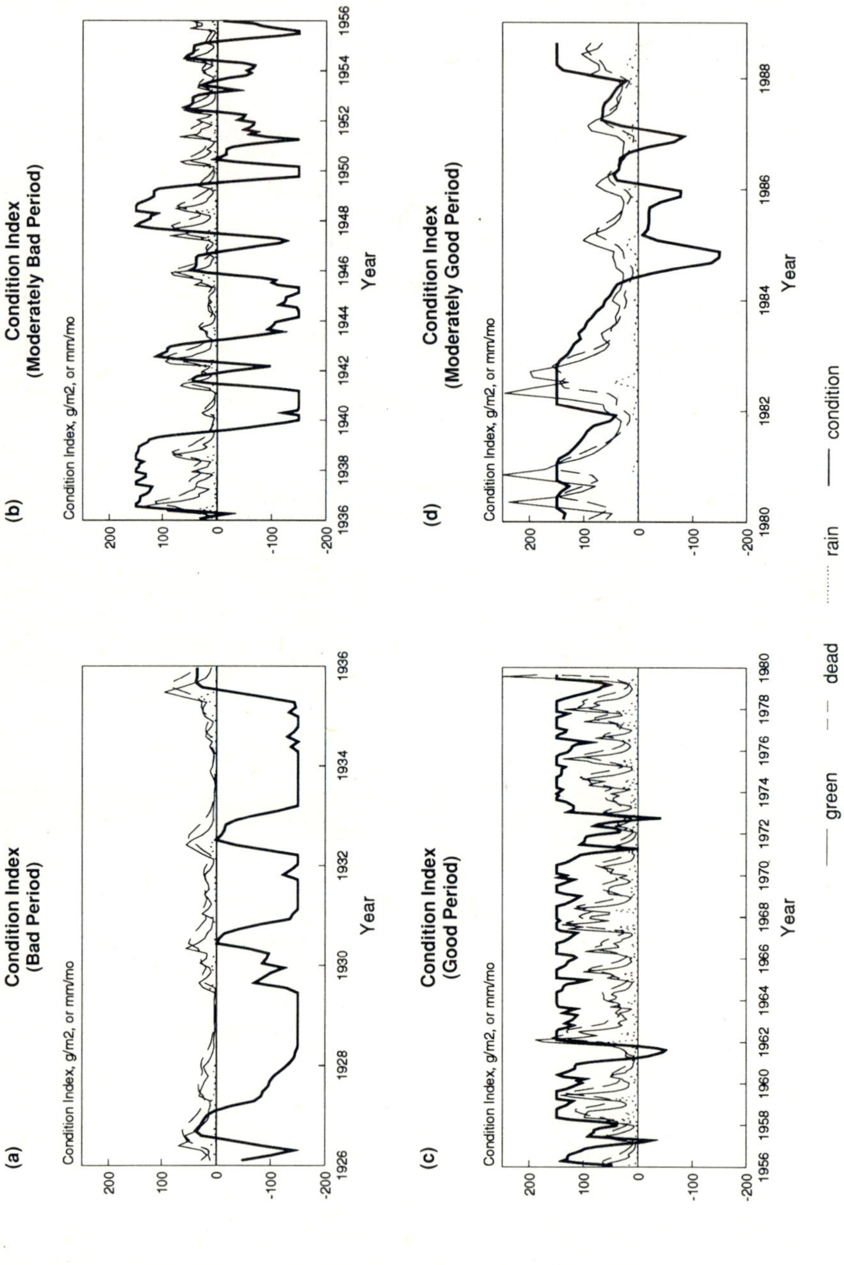

We believe that this simulation study tends to support the hypothesis that the STEP study area demonstrates non-equilibrial dynamics, characterised by large variations in primary production from one year to the next, and volatile changes in livestock numbers. Some of our colleagues grudgingly allow that we are probably correct about Turkana, but suspect that it is an unusual ecosystem, demonstrating idiosyncratic dynamics; they suggest that most other ecosystems might operate in a more normal (equilibrial) fashion. This may be true, but only if other arid ecosystems respond differently to climate variation from the way the STEP region does, because comparable patterns of climate variability are prevalent throughout the Sahel and in the arid parts of southern Africa as well as in some dry temperate zones (Nicholson, 1989).

We expect that the dynamics of the STEP study area are, like those presented in the results of Caughley, et al. (1987), the rule rather than the exception for arid ecosystems. If this is correct, a crucial question to ask is: Under what climatic conditions are these non-equilibrial dynamics likely to occur? Caughley et al. (1987) suggest that CVs of annual rainfall greater than 30% will engender these sorts of dynamics. Coppock's chapter in this book suggests that a threshold for non-equilibrial dynamic behaviour may occur between 300-400 mm rainfall per annum. These may be good estimates for the dry tropics, but what about dry temperate zones where diverse patterns of dynamic behaviour also occur (Weins, 1977, 1984)? What about dry, cold regions where grazing systems are subject to blizzards rather than or in addition to droughts (Goldstein et al., 1990)? Will global climate change increase climate variability and the prevalence of non-equilibrial dynamics? Finally, we need to develop a better idea of how to manage rangelands and how to implement range and livestock development programmes under these conditions.

Speculations on the management of non-equilibrial grazing systems

Standard range management procedures are based on Clementsian succession theory, an equilibrial concept, and are implemented through efforts to maintain a desirable range condition class (Lauenroth and Laycock, 1989). The applicability of this concept to dry rangelands has long been questioned (i.e., Hyder et al., 1966) both on a practical and on a theoretical basis. Scientists working in the dry tropics have begun to rethink the theoretical basis for range management and to suggest alternative concepts and practices. Some generalities seem to be emerging:

(i) Secondary successional sequences are poorly understood for many dry rangelands and in some cases may not exist. It is impractical therefore to use hypothetical successional sequences as management guidelines.

(ii) Plant production and community structure seem to be largely under the control of climate in the arid tropics, although herbivores do modify these attributes. However, the herbivores and the herder/managers are often in a position of responding to externally driven change in the grazing system, rather than manipulating and controlling the system themselves.

(iii) Carrying capacities probably have some utility in highly dynamic systems, but as an abstraction rather than a goal. This is because the capacity is likely to change rapidly both within and between years. Increasing herbivore numbers to track ephemeral upward trends in forage dynamics is difficult in commercial systems and near impossible in subsistence systems. Sustained downward trends in forage dynamics require a rapid response on the part of the herder/manager to avoid herbivore losses.

Taking these kinds of concerns into consideration, Sandford (1983a) and Westoby et al. (1989) have proposed that the operative posture for the management of dry rangelands and livestock is one of opportunism. This is also a characteristic of subsistence pastoralist strategies. In the most direct sense, opportunism means being prepared to respond to the opportunities and difficulties which present themselves. From the viewpoint of development and management, this suggests a strategy which allows for high mobility and rapid de-stocking and re-stocking or other such tactics for dealing with uncontrolled changes in plant community structure and biomass.

One aspect of opportunism which is prevalent in arid subsistence pastoral systems is the reliance on external resources. Just as these systems are driven by externalities (the vagaries of the climate), so they depend on externalities for persistence. The Turkana, for example, obtain external resources during periods of drought stress by taking their animals to regions not grazed during good rainfall years (McCabe, 1987b). This often includes invading the traditional ranges of other tribes in Uganda or Sudan, either by permission or through force of arms. The other major mode of acquiring external resources has traditionally been through livestock raiding. Raiding was prevalent in southern Turkana in the early to mid-1980s, and still occurs in the northern part of the District. In the arid Sahel, pastoralists like the WoDaabe and Fulani have traditionally relied on the acquisition of external resources through trade relationships with farmers from the more mesic portions of the region. Disruption of these trade relationships has meant a serious hardship for these pastoralists over the last two decades (Dupire, 1962; White, 1986).

Trading and raiding have always been important strategies for many pastoral societies in arid lands around the world. This is at least partly because persistence in non-equilibrial systems is difficult; the optimal strategy for pastoral persistence may thus be routinely to exploit the non-equilibrial ecosystem during 'good times' and to obtain external resources when the need arises. If this is so, then the most important single development procedure to promote persistence may be to solidify the connections between these pastoral ecosystems and the sources and sinks of external resources. Nominally, this means facilitating the inflows and outflows of goods and livestock through trade or marketing systems and external economies which can absorb and dispense products to and from pastoral ecosystems as they fluctuate over time.

Development procedures which may be useful and appropriate in equilibrial systems will often be counterproductive and destabilising in non-equilibrial systems. Such destabilising practices would include those which limit the pastoralists' ability to obtain external resources (e.g., confining pastoralists to relatively small ranches or other such schemes) as well as other procedures derived from traditional range and livestock management.

The information presented in this chapter would appear to demonstrate firstly that both equilibrial and non-equilibrial dynamics occur in real-world grazing systems, and secondly, that appropriate management and development interventions are different for equilibrial and non-equilibrial ecosystems. It necessarily follows that scientists, managers, and developers must understand which kind of system they are working with if their efforts are to be successful rather than counterproductive.

3
Vegetation and Pastoral Dynamics in the Southern Ethiopian Rangelands: Implications for Theory and Management*

D. Layne Coppock

Introduction

There is much debate over the contemporary effects of pastoralists and their livestock on the sustainable productivity of drier environments in sub-Saharan Africa. The central aspect of this debate is the definition of the degree to which climate or consumers influence environmental trends. One view is that consumers reach densities that degrade environments from a previous condition of equilibrium (Lamprey, 1983). Another is that the dynamics of pastoral systems are non-equilibrial and primarily dictated by variability in rainfall (Ellis and Swift, 1988). The primary objective of this paper is to review research on aspects of vegetation change and pastoral population dynamics in the southern Ethiopian rangelands during the 1980s and relate these findings to the debate on equilibrial versus non-equilibrial patterns of pastoral ecosystem dynamics.

The study

Study area

The Borana Plateau area is mostly semi-arid with a bimodal rainfall regime that supports perennial savanna communities (see details below). Cattle-rearing Borana pastoralists reside in encampments which may remain in the same location for up to eight years (Cossins and Upton, 1987). Families are thus semi-sedentary but herds can be quite mobile in their search for water

---

* This paper consists of excerpts from the ecology and drought-impact chapters in the following systems study, which comprises the work of over 30 contributors: Coppock, D.L. (in prep.) *The Borana Plateau of Southern Ethiopia: Synthesis of Pastoral Research, Development and Change*. International Livestock Centre for Africa, Addis Ababa. Funding for this research was provided through The International Livestock Centre for Africa (ILCA), the Third Livestock Development Project (TLDP), and the Fourth Livestock Development Project (FLDP) which supported the student training activity during 1987-90. The help of numerous other staff of ILCA, TLDP and CARE-Ethiopia in carrying out various aspects of this work is greatly appreciated. The author also thanks P. N. de Leeuw (ILCA) for his helpful comments on an earler version of this manuscript.

and grazing in dry periods. One major focus of the production system and of pastoral social and organisational life is the system of deep wells, which provide water in dry seasons and form the basis for some 30 traditional grazing associations (*madda*) in the study area. As one crosses the outermost boundaries of the system large ponds and temporary wells in river beds become more important for dry-season watering, and other pastoral and agropastoral groups (i.e. Gabra, Garri, Guji, Somali, Burgi, etc.) become more common. The distribution of encampments on the central plateau is most easily visualised as occurring in rings which are formed within a 15km radius of a central well area, and which serve to optimise access to well water and grazing. The distribution of resources in the system thus serves to concentrate the effects of humans and livestock. More details are available in Cossins and Upton (1987, 1988a,b).

Study objectives and methods

Research on plant ecology reported here was primarily conducted during 1982-6 (Billé, n.d. 1982, 1985; Billé and Assefa Eshete, 1983a,b; 1984; Billé et al., 1983; Billé and Corra, 1986a,b; Assefa Eshete et al., 1986). One objective of this work was to formulate some hypotheses linking livestock pressure to compositional dynamics in the plant communities and conduct some preliminary tests.

In total the work focused on twelve 400km^2 blocks selected on the basis of ecological and development criteria (Billé, 1982). Analysis of the blocks was designed to compare environmental trends by contrasting frequency distributions of grazing intensity, degree of erosion, grass and woody cover, and density and age structures of woody populations from extensive, rapid sampling.

Study sites

Two of the twelve study blocks were located in the arid zone, eight in the lower semi-arid zone, and two in the upper semi-arid zone. The arid zone (19% of the study area) occurs at <1200 metres above sea level (masl), with an annual rainfall of <500 mm, while the upper semi-arid zone (22% of the study area) occurs between 1500-1700 masl with an annual rainfall of 600-700 mm. The lower semi-arid zone (40% of the study area) occurs between the arid and upper semi-arid zones.

A further 7% of the study area was in the sub-humid zone, which typically occurs at over 1600 masl and has 700-1200mm of annual rainfall. Reference will also be made to observations from work in the sub-humid zone that did not involve analysis of blocks (Assefa Eshete et al., 1986; Pratt, 1987a,b).

The arid and lower semi-arid zones occur on a diverse assortment of granitic and volcanic soils and their mixtures, and in many instances shallow soils contribute to an edaphic source of aridity in terms of soil moisture storage. Basement-complex soils tend to dominate more at higher elevations. Vertisol bottomlands make up the remaining 12% of the study area (Assefa

Eshete et al., 1986) but are not a major focus of this paper. More details on the regional ecology of the Borana Plateau are available in Assefa Eshete et al. (1986).

Blocks were compared that varied in the historical intensity of pastoral use and thus would serve as natural treatments. Studies emphasised in this paper compared:

(i) four blocks in the lower semi-arid zone [Dass, Goff, Melbana, and Web (Fig. 3.1)] that represented various stages of pastoral impact in the deep-wells area (Billé et al, 1983);

(ii) two blocks in the lower and upper semi-arid zones [Medecho and Did Hara, respectively (Fig. 3.1)] that varied in recent duration of pastoral use due to immigration in response to recent water development at Did Hara or emigration from Medecho (Billé and Assefa Eshete, 1983b);

(iii) data from one block at Gololcha (1,600-1,900 masl; Fig. 3.1) at the boundary between the upper semi-arid and sub-humid zones where development of a large, permanent pond has recently allowed a high degree of pastoral sedentarisation (Billé and Assefa Eshete, 1984) with 52 cattle per km^2 recorded in 1982 (Milligan, 1983a: 37). The annual rainfall at Gololcha is on the order of 750 mm (Coppock, in press).

The four semi-arid blocks studied by Billé et al. (1983) represented regions that had been used by pastoralists for centuries. All four regions were assessed to be generally similar in terms of density of well groups, soils, landscape, altitude, and potential vegetative physiognomy. Aerial surveys during June 1982 indicated that the average stocking rate for the four regions was 24.6 cattle per km^2 (Milligan, 1983a: 37). Recent grazing history varied as Goff, located to the East, was abandoned by pastoralists for a few years during the late 1970s because of security problems but was reoccupied by 1980. In contrast, the other regions had apparently been continuously occupied during the past few decades.

In the second study at Medecho and Did Hara, these regions were regarded as generally similar in terms of cattle density (average of 24 per km^2; Milligan, 1983a: 37), annual rainfall (550 to 650 mm), soil parent materials (mixtures of volcanics and granitic materials), and density of water points. They principally differed, however, in recent intensity of use. Over the previous six years (1976-82) Did Hara had endured a doubling of human population in response to the construction of permanent ponds that opened the area to sedentary encampments (some of which began to resemble large villages). Medecho, in contrast, had been occupied for centuries but was in a state of population decline by the early 1980s (some of the residents may have moved 60 km North to Did Hara). Evidence for the different population dynamics was inferred (in part) from variation in the proportions of occupied versus unoccupied encampment sites. There will always be some unoccupied sites because encampments move every five to eight years (Cossins and Upton,

Figure 3.1: Schematic diagram of 12 400km² vegetation analysis sites in the Southern Ethiopian rangelands

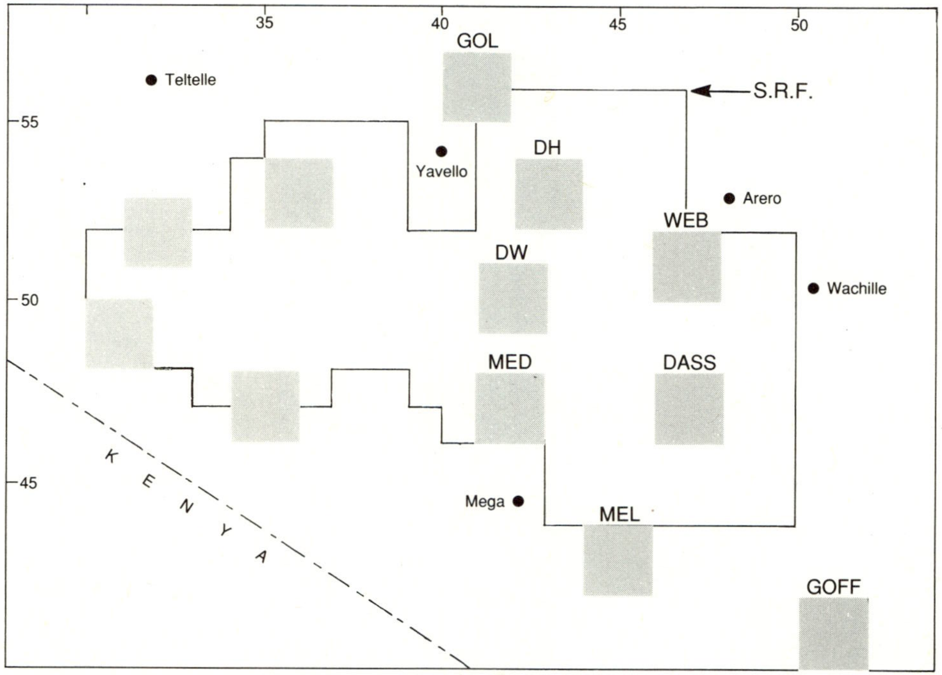

Notes: Where DASS=DASS; DH=Did Hara; DW = Demballa Wachu; GOFF= GOFF; GOL= Gololcha; MED = Mededcho; MEL = Melbana; and WEB = Web. Towns are denoted as dots. The solid line marked S.R.F. denotes the principal boundary for systematic reconnaisance flights for population estimation for livestock and people during 1982-86.

Source: JEPSS (1984)

1987). However, of 139 sites in Did Hara 39% were currently occupied in 1982-83, while at Medecho only 14% of 184 sites were occupied (Billé and Assefa Eshete, 1983b:30-31). Year-round access to permanent ponds in Did Hara probably also encouraged less mobile grazing strategies compared to Medecho. Cattle in Medecho radiate out in wet seasons to access forage near ephemeral ponds and return to graze closer to the central wells in dry seasons.

Ecological analyses

Blocks were initially mapped, using aerial photographs from 1965 and 1967 at a scale of 1:50,000, to delineate landscape facets with reference to previous surveys by AGROTEC/CRG/SEDES (1974a,b,c). Ecological units of similar vegetation and soil features were then identified using two colour composite prints from Landsat (ERTS E 2368 070 20501 in bands five and seven of 25 January 1976) and transcribed to a topographic map at a scale of 1:250,000. Ecological units were mapped and described further using ground-truthing during 1982-3. The number of ecological units per block ranged from 4 to 8. Detailed maps of each block may be found in Billé and Assefa Eshete (1983a,b; 1984) and Billé et al (1983).

Quantification of environmental variables in each block was based on an ocular survey method developed by Billé (n.d.) which employed up to 50 sample points per block, located along a driving route at 1km intervals. The route was designed to maximise observation of dominant ecological units in each block. At each sampling point a 1 ha. site was described in terms of:

(i) percent canopy cover for the herbaceous and woody layers, estimated into one of five classes;

(ii) density (no/ha) of woody plants;

(iii) one of five scores for topographic class, slope, degree of erosion and grazing pressure (Billé, 1982).

Random sampling of woody plants was used to obtain age structures of populations. Age class was estimated by measuring circumferences of single or central boles and placing them into one of seven categories from 1 to 70 cm. Plants with a bole circumference of 20 cm. were considered to be at least 70 years old (Billé and Assefa Eshete, 1983a:7).

Pastoral dynamics and perceptions of change

Effects of drought on cattle mortality were studied by Donaldson (1986) in a study based at five Borana encampments in the lower semi-arid zone where the fate of over 3,000 cattle was monitored for 18 months during 1983-4. General descriptions of drought effects are described in Cossins and Upton (1988a) and include notes on the incidence of human mortality and morbidity. Sex- and age-specific mortality rates, milk production and reproductive parameters were incorporated into a herd-growth model by Cossins and

Upton (1988a) to predict the time for the cow population and milk production to return to pre-drought levels. Coppock (in press) developed an empirical model based on the same cattle demographic data which also included density dependent production effects as inferred from producer interviews and projected population dynamics and food-energy deficits for the period 1959-2006. The model assumed growth in the human population at 2.5% per annum as inferred from recent demographic research (B. Lindtjørn, University of Bergen, pers. comm.).

Other surveys of human dynamics during drought were conducted for 90 households in the upper semi-arid zone between Did Hara and Gololcha by Webb and Coppock (unpublished data) in 1987 and 1989 by relying on recall of family and herd sizes before, during, and after the 1983-4 drought. Another survey on pastoral perceptions of cattle and human population dynamics and ecosystem change was conducted in 1990 based on in-depth interviews of 30 distinguished leaders of Borana society who resided in six *madda* (Coppock in press).

Results

Long-term effects in four lower semi-arid regions

Highlights of the results are presented here. Details may be found elsewhere (Billé and Assefa Eshete, 1983b; Billé et al., 1983, Billé and Assefa Eshete, 1984).

Based on an average of 43 sample points per block, the canopy cover of herbaceous plants was similar among blocks and ranged from an average of 20% for Web, Goff, and Dass to 29 % for Melbana. Blocks were more variable, however, in terms of woody cover. Goff had a woody canopy cover of 70%, while the other three blocks were similar and averaged 45%. This difference in woody cover represented variation in the density and size of woody plants. Size-class distributions for woody populations are shown in Fig. 3.2a-d.

These data were interpreted to show relatively mature and stable woody populations in Melbana, more balanced populations in Dass, and younger populations in Web and Goff. Inspection of areas under the curves in Fig. 3.2a-d indicates that the Web populations had only 22% of individuals with bole circumferences over 20 cm (or over 70 years old), while Goff had 24%, Dass 50%, and Melbana 78%. At both Web and Goff woody plants were considered to be reaching maximum densities (800 per ha). Billé et al. (1983:24) also noted that despite the general similarity of population curves for Web and Goff, there were important differences in the number of very old trees (i.e. those having a bole circumference over 30 cm). Less than 10% of the trees were in this category at Web, but this increased to about 30% (260 per ha) in Goff (Fig. 3.2c,d).

Frequency distributions for optimum and observed intensities of grazing and erosion for the four blocks are shown in Fig. 3.3a-c, and optimum distributions should be compared with observed to visualise trends. Current grazing pressure was skewed towards higher levels for Web and lower levels for Melbana (Fig. 3.3a). Erosion appeared more intense for Goff and less so for Melbana (Fig. 3.3c).

Figure 3.2: Frequency distribution for stem circumferences for sample populations of trees at four sites in the Southern Ethiopian rangelands

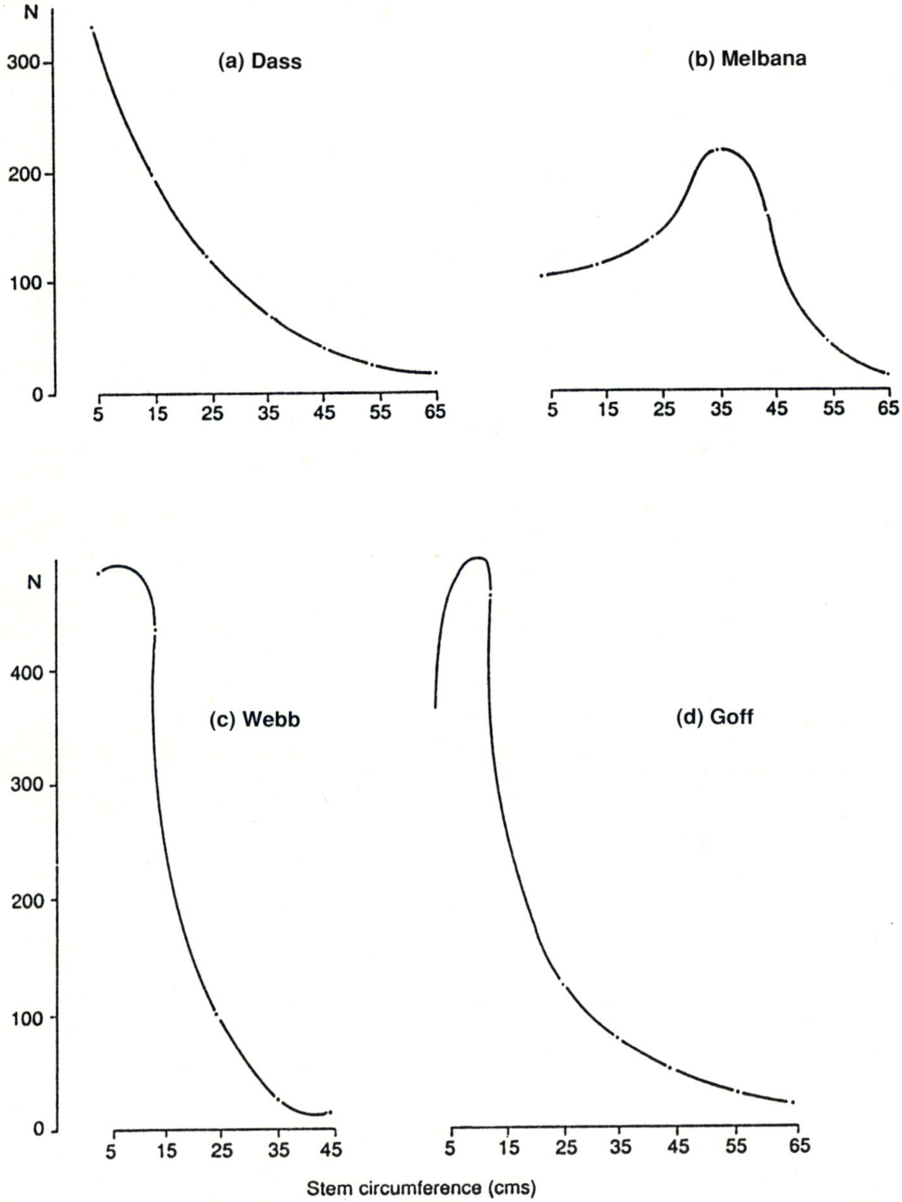

Source: Billé et al. (1983)

Figure 3.3: Frequency distribution for grazing pressure and erosion intensity, at four sites in the Southern Ethiopian rangelands

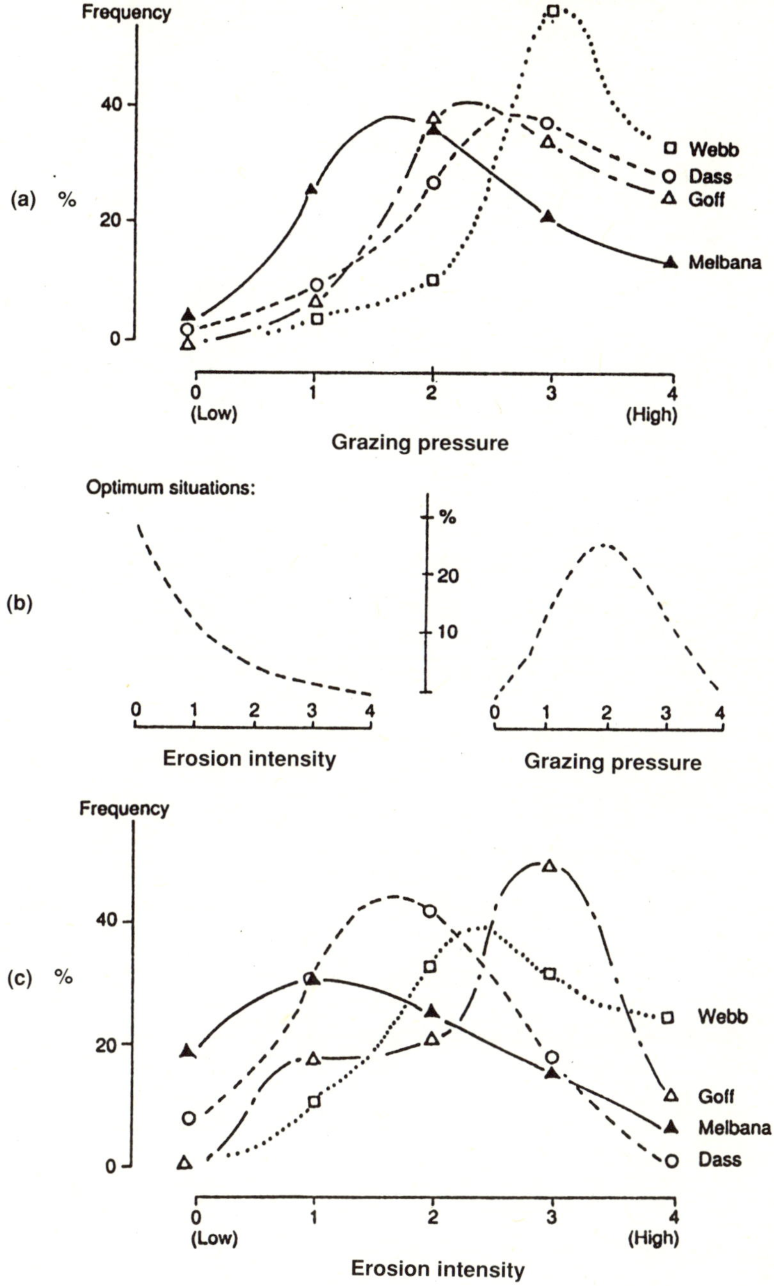

Source: Billé et al. (1983)

Billé et al. (1983) speculated that the data made sense if the four blocks were regarded as different stages in a historical cycle of heavy grazing, woody encroachment, reduced grazing pressure, gradual regeneration of the herbaceous layer, and subsequent decline in the recruitment of woody plants, as follows:

(i) Dass represented the initial stage of impact in which a high intensity of grazing had been occurring for a relatively short period of time. Some grazing-induced erosion had occurred, but the scattered woody population had not yet been affected in terms of age structure or density;

(ii) Web represented the next stage of impact in which the incidence of erosion increased and the establishment of young woody plants had occurred;

(iii) Melbana represented the third stage. Woody plants had increased here many years ago and the pastoralists subsequently abandoned the region (perhaps due, in part, to the heavy utilisation of grasses). The herbaceous layer then recovered in many places, establishment of more woody plants was curtailed, and the woody population stabilised and matured. The recovery of the grass layer probably started in the 1960s and was still in progress in the early 1980s;

(iv) Goff was hypothesised to be a variant of the third stage. It had apparently been heavily grazed for many decades and a high degree of erosion was noted to have markedly changed the landscape. The mixture of very young and very old trees suggested that successive waves of cattle grazing had not permitted the recovery of the grass layer as observed in Melbana.

Effects in one lower semi-arid and two upper semi-arid regions

Based on an average of 39 sample points per block, blocks at Medecho and Did Hara were similar in terms of herbaceous canopy cover (28-32%), but less so in terms of woody plant density (460 per ha. at Medecho versus 640 per ha. at Did Hara). However, the most interesting differences were in terms of age structure of woody populations. Did Hara had a younger *Commiphora*-dominated community with 64% of the population having a bole circumference below 10 cm, while the same size class at Medecho comprised only 25% of its populations (not illustrated). Conversely, Medecho had a higher percentage (26%) of very old trees (bole circumferences over 30 cm) compared to Did Hara (2%). Both blocks were heavily utilised by cattle at the time of the study, with grazing pressure rated as high to very high for an average of 65% of the sample points in both blocks (Billé and Assefa Eshete, 1983b:25).

The impacts of recent pond development (especially Beke Pond) on vegetation in Gololcha in the upper semi-arid zone was reported by Billé and Assefa Eshete (1984:22-7). Based on 47 sample points, the average density of woody plants in this block was estimated to be 1,170 per ha. Age structure of these populations was not assessed because of the density and complexity of the vegetation (Billé and Assefa Eshete, 1984:23). Gololcha was thus regarded as threatened by severe encroachment from species such as *Acacia brevispica* and *Commiphora* spp. (*ibid*.:23). Erosion and grazing intensity appeared to be inter-related and both were ranked as high to very high at over half of the sample points, the most of any block. Marked erosion occurred when grass cover was less than 20% (*ibid*.:26). Compared to other blocks, the pronounced impact in Gololcha was probably due to a combination of a higher density of livestock (see above), a reduced mobility of livestock, higher rainfall that favoured a faster establishment of woody vegetation, and a greater degree of hilliness that predisposes the area to cattle-induced erosion (ibid; Coppock, pers. obs.).

It was speculated that the environmental impacts of livestock at Medecho, Did Hara, and Gololcha illustrated the same principles observed at Dass, Web, Melbana, and Goff (discussed earlier). Recent and heavy grazing pressure in Did Hara, which had previously been relatively undisturbed by livestock because of poor water access, elicited a new wave of woody encroachment. Heavy grazing at Medecho in the past was evident in the mature age structure of the woody populations. Gololcha, under the highest rainfall regime, exhibited a particularly strong tendency for woody encroachment from grazing perturbation.

In sum, the traditional pastoral exploitation patterns, under former lower population densities of humans and livestock, were hypothesised to consist of the following stages (Billé and Assefa Eshete, 1983b:32-34; Billé and Corra, 1986a):

(i) Opportunistic settlements would be maintained around a given well group until the area was overgrazed to the extent that cattle productivity was compromised. Besides heavy grazing, negative effects on the herbaceous layer could also accrue from a possible reduction in the availability of top-soil nutrients (due to nutrient redistribution from forage in grazing areas to corrals at settlements via livestock faeces), and from encroachment of woody vegetation which could limit herbaceous growth through competitive effects for light, moisture, and nutrients.

(ii) These sites would eventually be abandoned, and the grass layer would begin to recover in the absence of grazing. Recruitment in the woody population consequently slows down owing to increased competition from grasses, and the woody populations mature. To some degree grass recovery could be facilitated by the physical obstruction of woody plants which limit livestock access, and the annual leaf litter from deciduous woody plants that could help replenish top-soil nutrients. Nitrogen fixation by woody legumes may also provide important inputs. As the herbaceous layer recovers further, fuel loads would build up and set the stage for fires to thin the woody populations. A fertile, open savanna is thus re-established.

(iii) Pastoralists then recolonise the site in a cycle that could take from 60 to 100 years for completion.

Pastoral dynamics

In the study of cattle dynamics in the lower semi-arid zone during the 1983-4 drought, Donaldson (1986) as well as Cossins and Upton (1988a) reported that about 30% of all cattle died and 19% were sold to buy grain out of an original total of 3,117. The relative impact of the drought on mature cows (45% died) and calves (90% died) appeared to be higher, compared to other age and sex classes of cattle. Although mature male cattle had among the lowest death rates (22%), more males were sold to purchase grain. These results are described in detail in Donaldson (1986), Cossins and Upton (1988a), and Coppock (in press).

The net loss of mature cows would jeopardise human welfare and herd growth during the post-drought recovery period. Modelling the recovery of cattle herds was reported in Cossins and Upton (1988a). They predicted that in regions where cows had been reduced by 25%, a complete recovery would occur by 1992 (i.e. in eight years) given 'normal' weather, while recovery would occur by 1994 or 1998, respectively, in regions that experienced 35% or 50% losses of cows.

These recovery estimates may be too conservative, however. A livestock census carried out after the drought indicated that many herds had apparently recovered their pre-drought numbers by 1988 (Solomon Desta, TLDP, unpubl. data). The discrepancy may be due to calving rates under conditions of good or average rainfall. Cossins and Upton (1987) considered a calving rate of 75% for normal production conditions, but it is conceivable that rates may have been as high as 90%, given a reduced competition for forage in the post-drought population (Coppock, pers. obs.). Regardless, the fast recovery of the herd has also been noted by the Borana (see below) and indicates that the system can be extremely productive and resilient.

Cossins and Upton (1988a) noted that although hunger-induced morbidity was common among people during the drought, there were actually few deaths attributed to famine. Similarly, Coppock and Mulugeta Mamo (1985) found that only 1% of the people reportedly died from famine in 60 encampments in the same region.

In surveys of 90 households in the upper semi-arid zone, Webb and Coppock (unpublished data) found an average drought loss of 33% for cattle. No deaths of people from famine were reported. However, about 90% of the families grew in size during the drought because each had one child born. Immigration or emigration of people was very rare.

In more remote regions of the Borana Plateau where precipitation is lower and market access is poor, effects of drought on the human population may be more severe (B. Lindtjørn, University of Bergen, pers. comm.). However, in 1985 there were only two famine camps in the region and the total number of residents may not have exceeded 5,000, including migrants from Kenya and elsewhere in Ethiopia (Coppock, pers. obs.). The scale of human deprivation observed here was less than that observed for six other sites in a national drought-impact survey in Ethiopia (P. Webb, International Food Policy Research Institute, pers. comm.).

The survey of 30 leaders of Borana society conducted in late 1990 (Coppock, in press) indicated that, according to the pastoralists, the cattle herd had recovered from the drought and had even recently entered a phase of density-dependent production effects, including reductions in milk production and increased risk of cattle mortality even in years of average rainfall. This high-density phase would coincide with stocking rates of over 20 head per km^2. The existence of density dependent effects on cattle has been subsequently confirmed by die-offs of cattle after the survey was completed (C. Fütterknecht, CARE-Ethiopia, pers. comm.). The hypothesis that cattle have reached some saturation point is supported by reports that drought reserve areas are being encroached upon by settlements and that reciprocal grazing rights were temporarily suspended in some *madda* (Coppock in press). This trend is also verified by the Borana's recent demand to the government to help re-excavate old wells with machinery. This is probably largely due to the need to water more animals and to exploit additional grazing areas (Coppock, pers. obs.).

Results from the empirical model (Coppock, in press) suggest that drought perturbations and density-dependent interactions will exacerbate food-energy deficits that are driven by an increasingly unfavourable imbalance in the ratio of milk cows to people, under the assumption that growth in the human population is more density independent (Webb and Coppock, unpublished data). Annual food-energy deficits for people (i.e. not covered by milk) may have risen from a hypothetical baseline of zero in 1959 to 40% in 1983, and may climb to 64% by the turn of the century. This confirms that grain will be an increasingly important subsidy to the system (Cossins and Upton, 1987), and places the human population at higher risks because of their increased dependence on grain imports from the highlands. The people attempt to cope with these problems through cultivation and increased sales of dairy products, small ruminants, and cattle (Coppock in press).

Discussion

General impacts on range resources

In addition to the dynamics reported for the analyses of blocks, in more extensive surveys it was later found that woody encroachment has occurred throughout the study area. Aerial surveys revealed that more than 20% (or 3,000 km^2) of the study area has over 40% woody canopy cover, concentrated in the eastern and central regions where cattle densities are highest (Cossins and Upton, 1988b:258). Such patterns are not unusual in semi-arid East Africa. Increases in woody vegetation have been reported elsewhere as a response to grazing (Norton-Griffiths, 1979; Cumming, 1982; Sabiiti and Wein, 1988; à Tchie and Gakahu, 1989; Belsky, 1989; Msafiri and Pieper, 1989).

Livestock pressure has also apparently influenced grass cover and the incidence of soil erosion far beyond that reported for the blocks. At one extreme, nearly one-third of the study area was later assessed from extensive regional surveys as degraded and in need of grazing control (Assefa Eshete et al., 1986). This is important, given that degradation was defined by these investigators as instances where perennial plant cover had been 'substantially' reduced and erosion had occurred as sheets or gullies. It was speculated that these impacts had occurred over many years and were largely due to grazing during rainy seasons, with effects magnified on shallow soils (ibid:23-24).

Other observers in Borana have come to similar conclusions regarding pastoral impacts. During the early 1970s when the increase in human and livestock numbers was just beginning in the Did Hara region in response to pond development, consultants from AGROTEC/CRG/SEDES (1974a: 85-87) noted that woody encroachment was a risk there. They also stated that woody encroachment was already advanced in a number of other semi-arid regions. Reporting on a brief survey, Pratt (1987a:15, 17) noted that, given the favourable rainfall, natural succession would lead to a dominance of woody plants in most communities regardless of grazing. He recognised the enhanced soil erosion at higher elevations and speculated that this was caused by high stocking rates and trekking of cattle to up-country markets. He (ibid.:18) also reported extensive bushland and thickets below 1500 m elevation and hypothesised that the overall increase in woody plants appeared to be due to a gradual thickening of existing stands due to seedling recruitment rather than expansion of plants into new sites. He (ibid.: 17-18) concluded that, although woody encroachment was marked, the rangeland was still in 'reasonable' condition because patches of high-quality perennial grasses remained in heavily grazed sites that could respond well to adequate rainfall, and because of the similarity in grass cover within government ranches (managed at lower stocking rates) compared to adjacent communal rangeland. Pratt (ibid.) felt that priority attention should be given to the spectre of rapid resource deterioration at higher elevations where population pressure was most acute.

Finally, summarizing another survey, Hacker (1990:5-24) also confirmed a high incidence of soil erosion as well as the presence of young woody plants in 68% of 801 survey sites. He noted that erosion appeared to be greatest on slopes comprised of red (i.e. yellow-red Haplic Xerosols) soil where grass cover was reduced by grazing, but that the amount of grass cover was not strongly associated with the abundance of young woody plants. In contrast to the more favourable assessments three years earlier in the same region by Pratt (1987a), Hacker (1990) ranked 43% of his sample sites as having poor to very poor grazing condition and felt that the government ranches had markedly higher grass cover than areas outside (Note that the discrepancy between the range condition surveys of Pratt (1987a) and Hacker (1990) may be explained by growth in the cattle population during this period, as discussed in the previous section). From a comparison of 15 sample points near the town of Yavello using aerial photos of 1967 and 1984, Hacker (1990:27-37) estimated that woody canopy cover had increased only slightly (from 24.1 to 28.5%), and that cover had increased at only four and decreased at one site. He concluded that changes in woody cover were thus more local than general in nature.

In sum, the consensus is that soil erosion and woody encroachment have occurred to various degrees throughout the study area. While livestock pressure can probably be directly linked to soil erosion, mechanistic links of cattle to woody encroachment are less clear. The simplest explanation is that by reducing fuel loads, heavy grazing reduces fire risk, which in turn allows more woody seedlings to establish. This model could explain most of the observed dynamics, including those of Hacker (1990), and is probably increasingly relevant as one moves up the rainfall gradient. It is expected that the tendency for woody plants to proliferate under grazing perturbation is greatest in the upper semi-arid and sub-humid zones.

In addition, woody encroachment could also be facilitated if grazing modifies the competitive relations among grasses and woody seedlings for soil moisture and nutrients, an interaction which may be most pronounced on fine-textured (rather than sandy) soils (Walker, 1985; Knoop and Walker, 1985). Most of the upland soils on the Borana Plateau are relatively fine-textured, and not sandy (Kamara and Haque, 1988).

Another factor that has undoubtedly facilitated woody encroachment has been the national policy which has banned burning of grazing and agricultural lands since the mid-1970s. Prior to this the Borana burned the rangeland to help improve forage quality and to control woody plants and ticks (Coppock, 1990). This policy constraint has been noted by several other investigators (Billé, 1985; Pratt, 1987a:18; Hacker, 1990: 20). Despite the policy, however, it is notable that extensive woody encroachment pre-dated the ban (AGROTEC/CRG/SEDES, 1974a:85-87; Billé et al., 1983), which suggests that traditional use of fire was only partially effective in the control of woody vegetation.

The hypothesised cycle of grazing-induced vegetation dynamics

Billé and Assefa Eshete (1983b:34) noted the traditional importance of the hypothesised cycle of cattle patch-use and shifts in vegetation composition in maintaining the long-term, dynamic stability and diversity of the Borana ecosystem. Factors that could disrupt the cycle, such as higher human and cattle populations that re-use sites before grass recovery has taken place (as at Goff), or higher populations of browsing stock (e.g. goats) that prevent the establishment of woody vegetation, could have some negative consequences for the sustainability of pastoral production. The same would hold if fire were too frequent (Coppock, pers. obs.). This hypothesis illustrates the point that woody encroachment could be beneficial over the long term by forcing pastoralists to emigrate from over-grazed sites and by providing a recovery opportunity for abused sites through replenishment of soil fertility as well as physically obstructing access by cattle in some cases (Billé and Assefa Eshete, 1983b).

In a rigorous sense, the elements that underpin the hypothesis of a cycle of grazing-induced woody dynamics remain untested here. Observations are only preliminary and require further research. However, some of the key elements are supported by other work in Borana as well as research from similar systems elsewhere:

(i) Migration of pastoralists in response to local resource base condition

Observations that the Borana occupy areas until they are severely over-grazed are supported only by the cursory reports of Billé and Assefa Eshete (1983b) in Medecho. Pastoral response to advanced stages of woody encroachment can only be surmised from interviews. In a survey of 60 encampments, Coppock and Mulugeta Mamo (1985) reported that although the Borana regard initial and intermediate stages of woody encroachment as beneficial or at least inconsequential to their production system, once encroachment is advanced they perceive negative effects in terms of shading on grass production and provision of cover for predators. Other herdowners have said that, instead of attempting to manage woody encroachment, they would prefer to move (Coppock, unpubl. data). This is a logical response for households that have considerable labour constraints (Mulugeta Assefa, 1990), and because the benefits accruing to individuals from implementing difficult interventions such as bush clearing in communal grazing areas cover probably only a fraction of the costs.

Although the people are semi-settled, they are highly mobile if the need arises. Encampments typically change location once every five to eight years, and this is commonly related to the search for better grazing (Donaldson, 1986; Cossins and Upton, 1987). Herdowners have recently reported that today it is increasingly difficult to relocate to improved grazing sites because of the general increase in the numbers of people and cattle (Coppock in press). They also think they are more sedentary than in previous generations, and this is reportedly due to the pull of roads, markets, and permanent water development that encourage settlement (Menwyelet Atsedu, TLDP/ILCA, unpubl. data).

(ii) The role of livestock grazing and corraling in relation to nutrient redistribution

Heavy grazing could serve to concentrate a significant quantity of nutrients from surrounding foraging areas because cattle are corralled nightly at encampments. This has not been quantified, however. Tonnes of manure are neatly piled into small hills at each Borana encampment, and this often represents years of accumulation (Donaldson, 1986). Elsewhere in pastoral Africa it is well-known that corralling of stock confines animals for about 50% of the time and consequently serves to concentrate large quantities of nutrients (Jahnke, 1982:35; Powell, 1986; Coppock et al., 1988). In addition to nutrient transport, frequent defoliation of perennial grasses during growing seasons could serve to mine nutrients in the top soils of foraging areas by stimulating translocation (Botkin et al., 1981).

(iii) The role of woody encroachment in facilitating recovery of the herbaceous layer

If nutrient depletion in top soils is a factor in the reduction of herbaceous productivity under heavy grazing in Borana, there is other evidence that woody encroachment could have some beneficial effects on recovery of herbaceous vegetation by improving soil fertility. First, since woody seedlings can quickly grow deep roots (Gates and Brown, 1988; Solomon Kebede, 1988; Tamene Yigezu, 1990) they could probably establish and tolerate nutrient-depleted top soils to a higher degree than shallower-rooted grasses.

Secondly, woody plants have been shown to have a potentially significant role in nutrient turnover and replenishment of soil fertility in semi-arid savannas. Billé and Corra (1986a) made some initial calculations of nutrient budgets for sites on the central Borana Plateau and estimated that leaf litter from at least 200 woody plants per hectare could provide roughly 57% and 21% of the annual turnover for nitrogen and phosphorus, respectively.

Hatton and Smart (1984) described the effects of a 24-year exclusion of wild herbivores from a Ugandan savanna. Under natural conditions, elephants prevented the persistence of woody populations through their feeding and clearing activities. Once elephants were excluded, however, an increase in *Acacias* occurred and these extracted nutrients from the soil profile and deposited litter on the soil surface. Compared to unprotected sites, top soils of protected sites showed up to a five-fold increase in exchangeable cations, a 50% increase in nitrogen, and up to a 30% increase in organic matter. Similarly, Radwanski and Wickens (1967) in the Sudan and Radwanski (1969) in northern Nigeria found that *Acacia albida* and *Azadirachta indica*, respectively, improved the fertility of top soils. Bosch and Van Wyk (1970), Charley and West (1975), and Weltzin and Coughenour (1990) also found that woody plants improved soils and/or micro-climate for herbaceous plants. Dreyfus and Dommergues (1981) highlighted the potential role of nitrogen fixation by native woody legumes in improving soil fertility when colonising new sites.

It is important to note, however, that whether interactions among woody and herbaceous plants are positive, neutral, or negative depends on soil texture (Walker, 1985), soil fertility (P.N. de Leeuw, ILCA, pers. comm.), harshness of climate (Weltzin and Coughenour, 1990), density and maturity of the woody layer, and potential for species-specific competition based on the compatibility of the seeds, juveniles and mature forms of plants (Harper, 1977). Of particular importance for species-specific interactions are root morphology and distribution (Walker, 1985) and the degree to which woody crowns intercept light from the understory. Whether plants are adapted to drought and/or fire would also play important mediating roles.

Some examples of the diverse interactions among plants in Borana include the apparently positive or neutral effects of *Acacia tortilis* (Billé and Corra, 1986b) and *A. drepanolobium* (Tamene Yigezu, 1990) on the herbaceous layer, and the negative effects of *A. horrida* and *A. seyal* (Solomon Kebede, 1988). A major hypothesis in explaining these diverse interactions is variation in the root and crown morphologies of the woody plants. The most marked contrasts include those species with lateral roots, short boles, and low, spreading crowns (i.e. *A. horrida*) which may confer the most negative effects, and those having tap roots, tall boles, and elevated, open crowns (i.e. *A. drepanolobium*), which probably confer only minor effects.

Pastoral population dynamics in relation to vegetation

Evidence shown here indicates that the cattle herd may recover very quickly from drought, and that density-dependent factors may take hold soon thereafter. If it is assumed that: (i) a multi-year drought occurs once every 20 years (Cossins and Upton, 1988a), and (ii) the cattle herd takes five years to recover, there is at least a 15-year high-density phase in the cattle population in which density-dependent interactions occur. If a stocking rate in excess of 20 head per km^2 is associated with a capability to effect change in the vegetation under this level of rainfall (Pratt and Gwynne, 1977), then the high density phase is the time when responses such as woody encroachment are most likely to happen. The degree of woody establishment would depend on rainfall. If the rainfall is above-average during the high density phase, the cattle population will be maintained at high levels with a more pervasive effect on the environment. Seed germination of woody plants would also be favoured by the higher precipitation (Tamene Yigezu, 1989).

The relatively low frequency of drought on the Borana Plateau, in conjunction with a higher human population that partially drives livestock production through management, thus permits a cattle population that makes an impact on the environment in a series of episodic events (Coppock in press). Woody encroachment is the most pervasive measure of the costs of this mode of opportunistic animal production, particularly if woody vegetation is allowed to mature and negatively influence herbaceous productivity. If human population growth is relatively density-independent and high (i.e. 2.5% per year), grain subsidies need to be available at favourable terms-of-trade to offset the imbalance between people and the milk-cow component of the regional herd. If not, human welfare will become increasingly precarious (Coppock in press).

Equilibrial versus non-equilibrial dynamics of pastoral systems

The patterns of resource impact and population dynamics observed in Borana largely conform to the mainstream view of pastoral ecology and economic change in semi-arid environments (Coppock in press). A similar scenario has been previously illustrated for Kenya's Maasailand (Meadows and White, 1979; Evangelou, 1984; Nestel, 1985; Grandin, 1987). The sequence consists of several general stages typified by increased population pressure. These include: (i) the pastoral population becomes increasingly confined to a fixed territory; (ii) numbers of cattle and people gradually increase as a result of health interventions and/or water development; (iii) degradation of vegetation and soil becomes more pervasive; (iv) the production system is more susceptible to even modest fluctuations in annual rainfall; (v) droughts may have progressively larger negative effects on livestock and people, who have fewer animals per capita to sell to purchase grain; (vi) growth in the human population eventually exceeds the potential for milk to fulfil the dominant portion of human energy requirements; (vii) the pastoralists must then diversify their economy to include more home-grown and purchased grains, sell a higher percentage of their cattle and increase their marketing of dairy products and small ruminants; and (viii) there is increased emigration of young males to other sectors of the national economy, who then provide cash remittances which help subsidise those who remain in the pastoral system.

This logical pattern is based on at least the episodic occurrence of equilibrial forces, which drive a wide variety of biological, ecological, economic, and social dynamics on the Borana Plateau (Coppock in press). Even the cycle of grazing-induced vegetation shifts hypothesised by Billé et al. (1983) and Billé and Assefa Eshete (1983b) is indicative of a patch dynamic for resource use postulated by Lamprey (1983) as within an equilibrial paradigm of pastoral ecology.

In contrast to the equilibrium concept, other investigators have noted that some pastoral systems appear to be non-equilibrial (Horowitz and Little, 1987; Ellis and Swift, 1988). This paradigm focuses on the hypothesis that vegetation dynamics are largely controlled by large variability in rainfall, and that livestock are essentially along for 'a roller coaster ride'. As such, livestock do not attain the densities necessary to have significant effects on vegetation composition, nor on primary or secondary productivity.

It is contended here that these two views are compatible in East Africa, if a distinction is made between arid systems (having unimodal or weak bimodal delivery of an annual rainfall of <400 mm) and semi-arid pastoral systems (having a bimodal delivery of 400-800 mm of annual rainfall). One major difference between these systems is drought frequency, as variability in rainfall is inversely related to the long-term annual mean (Nicholson, 1980). In semi-arid Borana a multi-year drought frequency of once in 20 years (Cossins and Upton, 1988a) is roughly four times greater than the five years needed for the cattle population to have recovered from the 1983-4 drought (Solomon Desta, TLDP, unpubl. data; Coppock in press). In contrast, in arid South Turkana within Kenya, droughts may occur as often as once every five years (Ellis and Swift, 1988), which is roughly equivalent to the drought-recovery period needed for cattle and camels (McCabe, 1987a). It is thus not surprising that livestock population dynamics, and the prospects for livestock-induced impacts on resources, could markedly differ between these systems simply due to drought frequency alone.

Other factors to consider, however, are differences in soils, landscapes, and vegetation between Borana and South Turkana, some of which are also related to the rainfall regime. In Borana the silty soils are readily erodible on hilly landscapes, while much of the Turkana system consists of poorly differentiated sandy or volcanic substrates on level plains (Coppock, 1985) that would not be expected to exhibit structural changes even under high intensities of livestock use. In addition, while Borana is dominated by perennial grasses, South Turkana is dominated by annuals (Coppock, 1985). Annuals have several important features that would contribute to the severity of drought impacts and the resilience of an arid ecosystem. First, even a two-year drought can result in virtually no standing herbaceous biomass, as observed in 1981 on the sandy plains of South Turkana (Coppock, pers. obs.). Annuals persist in seed banks, and standing biomass is gradually lost either by weathering or grazing. This means that drought in South Turkana can have a devastating impact on cattle, irrespective of their pre-drought densities. During years of average rainfall, herbaceous productivity depends on variation in the seed bank. Upon receiving sufficient rainfall, annual grasses mature and set seed within six weeks (Coppock, 1985). The number of grazers needed to compromise this seed production within this time limit is very high, and in most years probably much higher than the existing population. Thus herbaceous production and cattle populations are relatively uncoupled to a significant degree. Rainfall, not consumers, largely dictates the composition of vegetation from year to year.

Perennial grasses in Borana provide a sharp contrast to these patterns. Although many cattle emigrated from the study area during the 1983-4 drought (Donaldson, 1986; Cossins and Upton, 1988a), the persistence of perennial grass cover sets the stage for a longer period of grazing interactions and density-dependent effects during drought, albeit at lower stocking rates. The greatest effects of cattle in mediating compositional changes in the vegetation probably occur during the high density phase of the cattle population in average rainfall years (as has occurred from 1989 to the present). To the extent that woody encroachment or soil erosion compromises herbaceous productivity, negative carry-over effects are anticipated for cattle from one high-density phase to the next (see Sandford (1983a) for a discussion of the significance of carry-over effects in relation to costs of opportunistic pastoral production). Perennial grasses would, however, rebound to some extent during each drought-recovery period, depending on rainfall. Rather than one continuous, readily discernible trend, patterns of resource degradation in Borana most probably consist of episodic events that vary in relation to whether rainfall is above- or below-average, which determines the duration and intensity of plant-animal interactions in the high-density phase (Coppock in press).

In sum, the dichotomy of arid and semi-arid systems in this analysis is supported by other theoretical frameworks. Noy-Meir (1973) noted that systems having less than 400 mm of annual rainfall are under abiotic control in relation to water; non-equilibrial dynamics of consumers should prevail in these instances (Ellis and Swift, 1988). The semi-arid Borana system, however, falls more within the realm of at least periodic biotic regulation with the frequent occurrence of equilibrial dynamics among primary producers and consumers (Norton-Griffiths, 1979; Cumming, 1982; Walker, 1985).

Conclusions

Under increasing population pressure in a semi-arid environment, the Borana pastoralists and their cattle have elicited dynamic changes in their environment, and the prognosis is that their effects may become more acute in the future. Impacts will be most pronounced in the regions of highest rainfall. Carrying capacity, successional trends, and density-dependent population interactions are viable concepts here. It is likely that, rather than offering conflicting views on the dynamics of the same type of pastoral systems, researchers espousing the dominance of biotic or abiotic control may be interpreting observations from functionally different systems. In Borana drought pulses confuse the interpretation of dynamics at any one point in time, but the fundamental pattern is more of an equilibrial rather than a non-equilibrial situation. Development efforts to ameliorate the negative effects of the pastoralists on the environment may initially focus on tactics such as bush control, range rehabilitation, and improving drought security, but these will only be temporary solutions in light of the driving force of human population growth. The ultimate development strategy for the future is policies and procedures which facilitate diversification and security in the traditional economy and provide the skills and choices for a segment of this society to pursue alternative lifestyles.

4

Why Are There So Many Animals? Cattle Population Dynamics in the Communal Areas of Zimbabwe*

Ian Scoones

Introduction

The question of why stocking rates in Zimbabwe's communal areas are so high has been asked for a long time (Jarvis and Erickson, 1987). Officials have assumed that the cattle populations have been 'exceeding ecological limits' and during the colonial period a series of destocking measures were imposed to bring the populations down to the 'carrying capacity' of the land (Scoones, 1990). This debate still continues and current policy still advocates destocking, regulation of grazing and grazing schemes (Government of Zimbabwe, 1986).

These policy measures assume that ecological carrying capacity is being exceeded, that there is some optimal equilibrium between populations and resources and that control measures will have a direct effect. This, in turn, assumes a particular type of population dynamics, based on equilibrium states and density dependent limitation of populations. In many pastoral systems, especially where rainfall is highly variable and has a major influence on primary and secondary productivity, these assumptions may not hold (Ellis and Swift, 1988). This has important implications for the form of management adopted.

This chapter explores the forms of population limitation and regulation in the cattle populations of the communal areas in the central-south of the country. The wetter areas of this zone are situated in the dry miombo woodland belt, with sandy soils where rainfall averages c. 800mm annually. The drier areas receive less than c. 600mm rainfall annually and are found in the heavier soil areas dominated by *Acacia* and *Colophospermum mopane* vegetation associations (Scoones, 1990).

---

* The research for this paper was undertaken while at the Renewable Resources Assessment Group, Imperial College, London and while a visiting research associate at the Department of Biological Sciences, University of Zimbabwe. Johnson Madyakuseni assisted with the study of cattle foraging behaviour and Abraham Mawere helped with the detailed study of sample population cattle. The Department of Veterinary Services and the Economics and Marketing Branch, Ministry of Lands, Agriculture and Rural Resettlement provided much of the quantitative time series data on livestock populations. The research was funded by the SERC/ESRC, University of London and the International Institute for Environment and Development (IIED).

Population dynamics

The confusion generated by the classic debate over density dependence and density-independent factors is clarified by differentiating between population limitation and regulation. Population limitation relates to factors that determine the absolute numbers of a population; these may be density dependent or independent factors. Population regulation relates to, by definition, only density dependent factors (Sinclair, 1989).

A population may follow the expectations of equilibrium theory, in which case a stable state, or possibly multiple stable states (May, 1977), is reached where an equilibrium between population and resource base is found. Density dependent regulatory factors then come into play. This has been the dominant 'model' for ecological theory since the early work of Nicholson (1933). However, non-equilibrium dynamics may be observed, where a population never reaches a level where density dependent regulatory factors act and populations are limited by largely density independent factors, such as rainfall. A range of dynamic behaviour, from stability to periodicity to chaos, may be observed, resulting from simple dynamic interactions (May, 1977). Non-equilibrium dynamics are less tractable in terms of simple deterministic models (Caswell, 1982), but may represent the dominant form of dynamics in dryland areas (Wiens, 1984), where rainfall has such an important role in savanna dynamics (Frost et al., 1986). In pastoral populations this may be particularly so (Ellis and Swift, 1988).

In reality, animal populations, including livestock populations, may be affected by a mix of factors, with density dependent regulation acting alongside the more dramatic impacts of density independent limitation.

It is not a trivial question to ask what is the expected mixture of factors of regulation and limitation, since the form of dynamics will greatly affect the type of policy intervention that will be relevant. An exploration of dynamics will also offer possible answers to the question: why are there so many animals?

There are various routes to understanding in more detail the dynamics of population regulation and limitation. A number of options are noted in the literature:

- life-table and key factor analysis (Varley and Gradwell, 1960).

- census data analysis: simple total population or age/sex structured models (Caughley, 1977).

- perturbation experiments: natural or experimentally induced disturbances in populations and observations of recovery dynamics (Sinclair, 1989).

This chapter will use the latter two options, with data from cattle populations derived from Veterinary Department records, for the period between 1923 and 1986 for six communal areas in central-southern Zimbabwe.

Patterns of cattle populations: change over the last century

Accurate data do not fully cover the whole of the last century, but acceptable inferences can be made of the situation before 1923 when a good coverage of dipping records began. A number of phases can be identified over this period (see Fig. 4.1). A series of major perturbations have occurred that have affected the pattern of the population trajectory. These have been due to disease, policy-directed destocking, war and drought. Despite the major impact of these events, populations have persisted. At each stage, following a perturbation, the cattle populations have recovered.

The phases of cattle population change are sketched out in a simplified form in Table 4.1. Populations in each of the areas did not follow exactly the same trajectory, but the broad pattern is common to all areas (for details of individual areas, see Scoones, 1990; Fig. 4.1).

The history of cattle populations in these areas is a history of perturbation experiments on a grand scale. Is it possible to discern from the available data whether populations have shown a pattern of density dependent regulation, with major density-independent reductions in population size intervening at various stages?

The evidence is equivocal. A good argument can be made that suggests that the population levels of the mid-late 1930s and the mid-late 1970s represent 'equilibrium' levels regulated by density dependent factors. In each case populations have recovered from previous perturbations and been maintained, if only for a few years, at the higher level. However, the populations reached during these two periods are not the same; the population peaks of the 1970s were higher.

It is clear there is no static 'equilibrium' level, since rainfall varies interannually and on a cyclical basis (Tyson, 1976). The 1970s were a 'wet period', while the 1930s were characterised by occasional, but not severe, droughts, the result being that available biomass, and so population 'carrying capacity', differed dramatically.

Another factor that confounds comparison between the two periods is the nature of land-use change in the area. In the 1930s, there was extensive grazing land and limited cropping land (perhaps 5-10%). By the 1970s cropping land accounted for 40% or more (Whitlow, 1979). The impacts of such change on livestock are uncertain and complex (Scoones, 1990).

An alternative hypothesis suggests that the patterns of population change can be explained simply by the impact of episodic events such as drought, disease and destocking. Periods between such events are sufficiently short not to result in populations reaching levels where density dependent regulation acts. Populations are thus simply limited by episodic, largely density independent factors. Whether drought impacts are density independent will be considered below, but it is unlikely that tick-borne diseases or rinderpest act completely independently of density.

The evidence for these alternative explanations of historical population dynamics is explored further in the next section.

Table 4.1: Cattle population change in southern Zimbabwe

pre-1896	Before the rinderpest pandemic of 1896, cattle populations in the area were probably high, competing with extensive wildlife populations. Records indicate that Ndebele herders were utilising much of the area as extensive grazing grounds for large herds (Cobbing, 1976).
1896	Rinderpest devastated cattle populations across the country (Mack, 1970), reducing populations to a fraction of their previous size. The removal of power from the Ndebele state and the seizure of cattle by the settlers also reduced cattle populations in the area.
1896-1945	This period showed a recovery of cattle populations to densities of 0.25-0.3 animals/ha in the study areas. Average growth rates between 1923 and 1935 ranged from 5 to 20% per annum in different areas. The establishment of dipping facilities from the early 1920s also reduced disease related mortalities.
1945-1960	The destocking policy, instituted by the colonial government in all the study areas, saw a major reduction in cattle populations. In some areas densities were reduced to 0.15 animals/ha, little more than the population density recorded in the early 1920s.
1961-1975	The abandonment of destocking allowed populations to recover again. The high rainfall levels of the 1970s provided plentiful fodder, and cattle populations recovered to levels not seen since the late 1930s. In some areas populations increased even more. Average growth rates between 1961 and 1978 were 3-7% per annum in different study areas.
1976-1979	The hostilities associated with the liberation war resulted in the abandonment of dipping services. This resulted in a large increase in tick borne diseases and another drop in population levels (Norval, 1978).
1980-1982	Following independence populations increased again, as dipping services were restored.
1983-1984	The major drought of this period resulted in large-scale mortality. Populations at the dip tanks in the Mazvihwa communal area were reduced by 60 - 80% (Scoones, 1990: 376).
1985-1986	A period of good years allowed a rapid recovery. Populations at the dip tanks in the Mazvihwa communal area increased by 18%-114% per year during this period (due to both natural increase and in-migration).
1987	A further drought resulted in more mortalities. Populations at the dip tanks in the Mazvihwa communal area were reduced by 8-40%.
1988-1990	A period of recovery has been observed again.

Figure 4.1: Livestock populations, 1923-1986

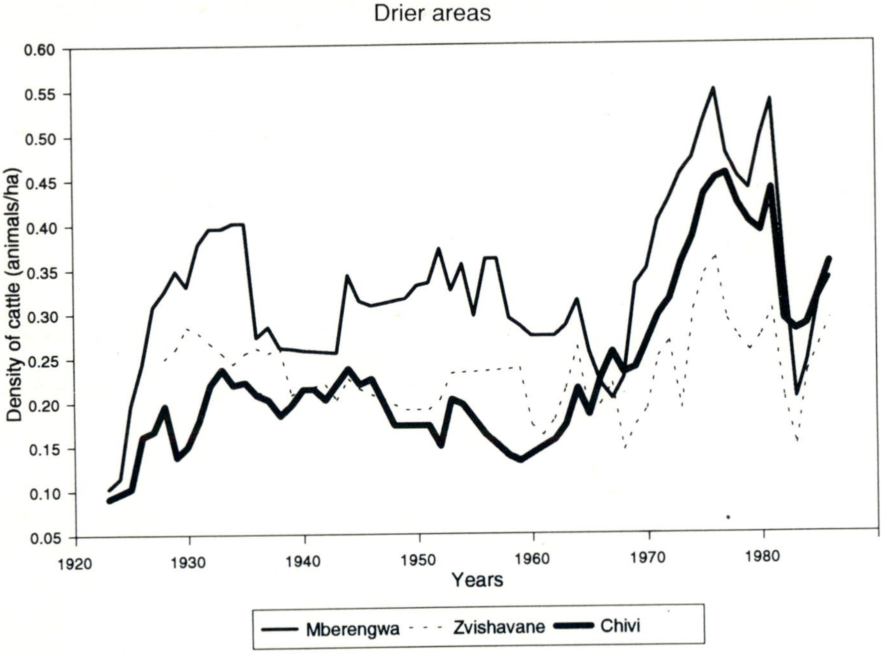

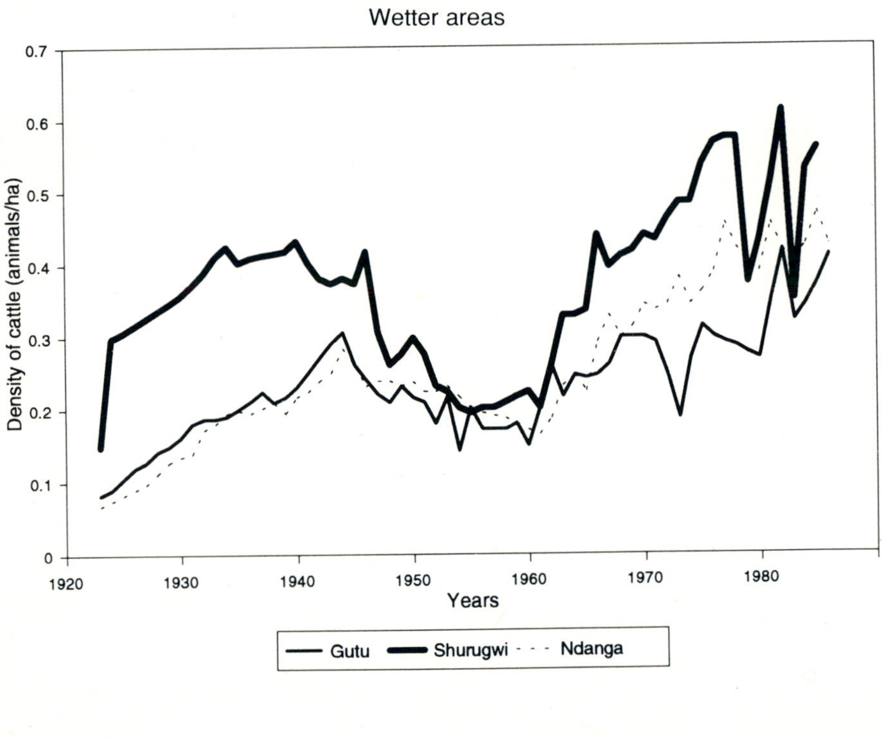

Population regulation or limitation: is there an 'equilibrium' population density?

Simple models that describe the basic dynamics of populations can be used to explore whether the observed patterns of population change demonstrate dynamics approximated by the models. Such models are not intended to replicate exactly the patterns observed, but are aimed at exploring underlying processes. However, it is difficult to identify underlying regulatory factors in the 'noisy' and heterogeneous state of the real world (Hassel et al., 1987, May, 1989a). May (1989b) notes:

> The basic problem is that once patchiness in distribution, environmental noise and chaotic dynamics all interweave, it can be difficult to disentangle the chaotic, density dependent signals from additive noise.

The census data on cattle population densities, and birth and death rates over time, are used in this section to investigate the degree to which density dependent processes are acting.

A simple logistic model is first used to explore the growth trajectory of the cattle populations in the six communal areas. A logistic model is a major simplification of reality. This is because herbivore-plant systems violate a number of the model's assumptions. Most importantly, the logistic model assumes that the food resource is inert and renewing and is unaffected by the herbivore. Interactive models of plant-herbivore dynamics are therefore more realistic, but they do require a range of parameter estimates which are unavailable for the present data (Caughley and Lawton, 1981).

There are innumerable variations of the logistic model. As they become more complex (incorporating further non-linear relationships, time delays, age-structured growth models, etc.), models become more data-hungry. For the present examination, simple models will be used; even these are known to describe complex dynamics (May, 1977).

The logistic growth model describes population growth (with harvest removals) according to the following difference equation:

$$N_{t+1} + N_t + H = rN_t - rN_t^2/K$$

where:

r = intrinsic rate of population growth
N_t = population at time t
N_{t+1} = population at time t+1
H = harvest+sales
K = ecological 'carrying capacity'

This model was investigated for the populations of the six study areas over different time-periods, to explore whether there were fundamental differences in growth dynamics in different periods (e.g. the period of early growth from 1923 to 1935, the period following destocking, etc., see Scoones, 1990). No major differences were found between different periods. The model fit to the data for the full time series for three communal areas is shown in Fig. 4.2.

Table 4.2: Logistic growth models for three communal areas, 1923-1986

Mberengwa, case 1 ($r^2 = 0.3$)

$N_{t+1} - N_t + H =$	$0.28N_t -$	$0.58N_t^2$
SE:	0.08	0.22

Zvishavane, case 2 ($r^2 = 0.34$)

$N_{t+1} - N_t + H =$	$0.5N_t -$	$1.72N_t^2$
SE:	0.13	0.52

Shurugwi, case 3 ($r^2 = 0.63$)

$N_{t+1} - N_t + H =$	$0.73N_t -$	$1.58N_t^2$
SE:	0.11	0.25

The results for three communal areas where the sales/harvest rate was regressed out are shown in Table 4.2. In each case $p < 0.01$ for the regression.

The models explain only a proportion of the observed variation. The real world is far more complex. However, the results do suggest that underlying dynamics have been partially accounted for. Additional factors that a more sophisticated model should include are: age and sex structure of the population, seasonal dimensions, spatial heterogeneity and multi-species use (Scoones, 1990).

The models suggest therefore that an overall 'carrying capacity' for the system would be as shown in Table 4.3.

The possibility of density dependent regulation can be investigated further by examining the relationship between recorded birth and death rates and density. A simple regression of death rate (%) against density of cattle population revealed no significant relationship. This might suggest that no regulation was occurring and the population trajectories examined by the simple growth models is best explained by stochastic factors. However, a more

Table 4.3: Estimates of 'carrying capacity' from logistic models

(densities with the sales/harvests regressed out; lower densities are reached when removals are considered).

Case 1	Mberengwa:	0.47 animals/ha
Case 2	Zvishavane:	0.29 animals/ha
Case 3	Shurugwi:	0.46 animals/ha

Figure 4.2: Simple population models for cattle in the communal areas, 1923-1986; 1 = Simple growth model 2 = Logistic growth model

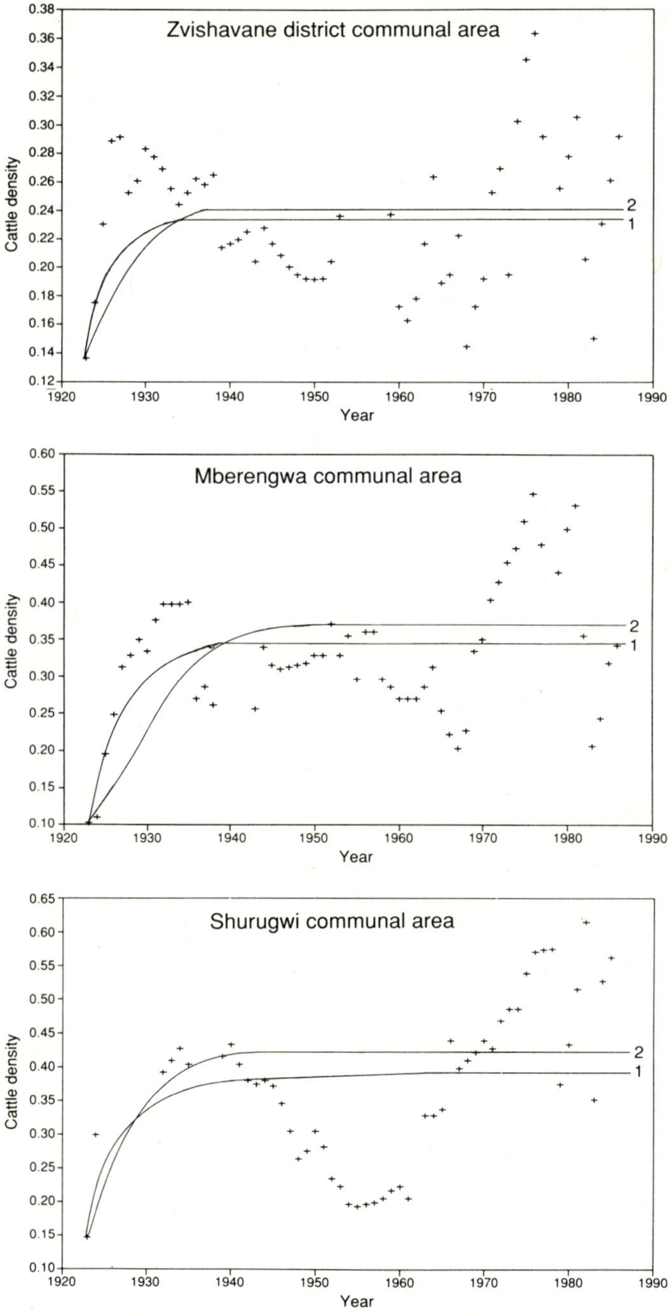

disaggregated investigation reveals more complex dynamics. An examination of the death rate data showed that mortalities were very high in some years (up to 25%), while mortalities in most years were around 2-4%. These 'stress years' showed up as outliers in the aggregate regression. Once these outlier stress years were removed from the data set, a different pattern emerged. Death rates during other years were found to be density dependent. Birth rates showed a similar pattern, but without the 'stress year' effect. In all years birth rate was negatively related to density.

The results of this analysis are shown in Fig. 4.3 and Table 4.4. Table 4.4 shows the regression results for the aggregate data for all study areas combined. An 'equilibrium' population level is reached when birth rates equal death rates, in the absence of immigration, emigration, sales, purchases or slaughters of animals. This may be calculated by an extrapolation of the regression line estimates relating birth/death rates against density. For the aggregate data, this 'equilibrium' point is at 0.82 animals/ha (see Fig. 4.3).

Table 4.4: Theoretical equilibrium levels

Death rates (non-stress years) = $7.51(N_t)$
$[r^2 = 0.61; N= 62; p < 0.001]$

Birth rates (all years) = $-21.7(N_t) + 24.01$
$[r^2 = 0.21; N = 61; p < 0.001]$

This density is higher than that estimated from the simple growth models (Table 4.3). Why is this? First, the estimate ignores the impact of 'stress years' on death rates. The impact of these will act to reduce the population dramatically as high mortalities occur in these years. Second, the estimate assumes a simple linear extrapolation of the regression relationships beyond the bounds of the data. However, it is likely that non-linear density dependent relationships operate in large mammals (Fowler, 1981, 1987). This would act to reduce the equilibrium point estimate (see dotted lines in Fig. 4.3).

It appears therefore that population size is likely to be limited by 'stress year' effects on mortality, reducing population levels below 'equilibrium' levels when such episodic perturbations act. However, as populations increase towards a notional 'equilibrium' or 'carrying capacity' level (which itself varies over time, according to rainfall conditions, among other factors), density dependent regulation does occur in relation to both birth and death rates, although the effect is weak compared to the impact of episodic stress years.

The impact of stress years is therefore likely to be very important in the overall dynamics of cattle populations in these semi-arid savannas. Do stress year mortalities act in a density dependent manner? Investigation of this question for the major stress years was limited by the size of the data set, but no relationship between major stress year mortality and population size in the previous year could be found. Most stress years are droughts; although a cold

Figure 4.3: Equililbrium density estimates

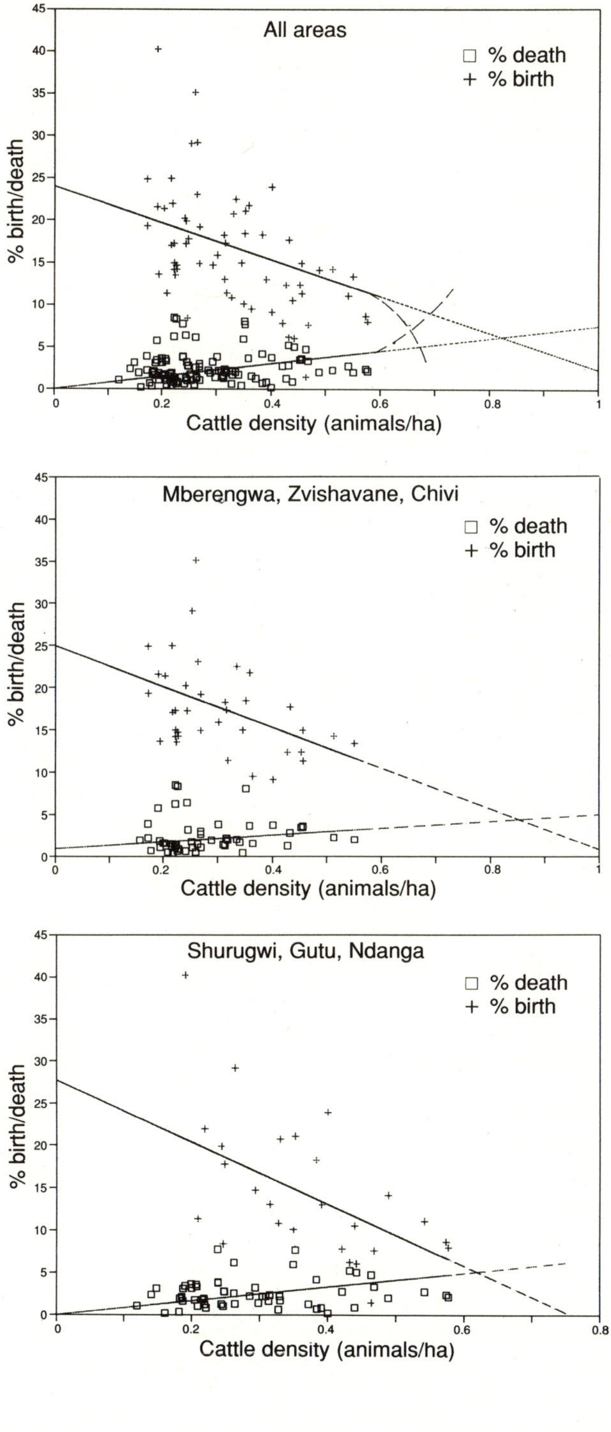

snap prior to the rains, as in 1968, can also have a significant impact. A sample of years, during the period 1923-86 when annual rainfall was 25% or more below the average, were investigated. Mortalities in these years equally showed no relationship with cattle density.

Major mortality levels having a significant impact on population size appear therefore to be density independent, the result of stochastic (or cyclical) variations in rainfall incidence. By contrast, mortality in non-stress years appears to have a density dependent impact. Again, this is not the whole story. Rainfall is also significantly correlated with mortality in these years, especially in the drier areas. Birth rates appear to behave in a more predictable density dependent manner; no relationships between rainfall (previous year, lagged or combined) and birth rate could be detected in the data set (Scoones, 1990).

This section has shown that the population dynamics of cattle populations are characterised by an interaction of non-equilibrium dynamics (major impacts during 'stress years' on cattle mortality) and equilibrium dynamics (underlying density dependence of birth and death rates). Non-equilibrium dynamics probably dominate, limiting populations to levels below potential 'equilibrium' levels.

However, these levels, even after the impact of 'stress year' effects, sales and harvesting etc., still represent high population densities; several times higher than official recommended carrying capacities. What is going on?

Why are there so many animals?

The estimates made of 'ecological carrying capacity' from the analysis of simple growth models have been shown in Table 4.3, with levels ranging from 0.29-0.47 animals/ha. The official 'carrying capacity' levels for these areas range between 0.14 livestock units (LU)/ha for the marginally wetter areas and 0.1 LU/ha for the drier areas (Agritex, n.d). If an average animal is assumed to be 0.66LU in the communal areas, then these figures translate to 0.21 and 0.15 animals/ha. The estimates made by the simple modelling exercise are between 1.9 and 3.1 times higher. Why?

The fundamental reason is that the two estimates relate to two different concepts. The estimates derived from the simple models refer to a rough estimate of 'ecological carrying capacity'. This varies with rainfall, but relates to the number of animals the land can sustain. This figure is less than the 'true' ecological carrying capacity which is defined as the population level where the rate of change is constant (i.e. the equilibrium level estimated in Fig. 4.3). The reasons for this disparity have already been considered. However, both these estimates of 'ecological carrying capacity' are significantly higher than those promulgated in official range management recommendations. These recommendations refer to the 'economic carrying capacity', derived from a beef production model of range management, and are irrelevant to the communal area system (see Scoones, 1989a for further discussion of these different measures in the context of livestock production in Zimbabwe).

This still does not answer the question: why are there so many animals? The data demonstrate that populations have persisted. Despite instability imposed by episodic reductions, the system is apparently resilient. What are the mechanisms for population persistence at the densities observed? It is necessary to ask:

- What are the management strategies used by herders to offset population-limiting mortalities - both density dependent factors that occur in 'normal' years and density independent factors that occur episodically in 'stress years'?

- What components of the forage resource and the livestock landscape are significant in population persistence? Are these areas the determinants of 'carrying capacity'?

These questions cannot be answered from the aggregated time series population data. It requires a local case study of a specific population to provide some clues. This case study was carried out in the Mazvihwa communal area of Zvishavane district from 1986 to 1988. This period included the single drought year of 1987. An examination of strategies during this year, and retrospective analysis of the situation during the drought of 1982-4, provide a detailed insight into patterns of communal area cattle population dynamics.

Local-level strategies: implications for population dynamics

Understanding longer-term dynamics requires a detailed investigation of behaviour patterns at the local level. Only with a combination of insights derived from different levels of investigation, from population dynamics to individual behaviour, can a more comprehensive picture of ecological processes be built up (Hassell and May, 1985).

The averaged dynamics of simple models are the result of interactions operating at different levels of a hierarchy. An examination of dynamics within patches existing at different scales is necessary to fully understand the patterns of population persistence and to answer the question: why are there so many animals?

Cattle forage at different spatial scales. At the widest scale, animals move between different savanna types. In the study area, movements between clay veld and sand veld are important (Scoones, 1993). Within savanna zones, movement up and down catenal sequences of different resource patches is observed, with a concentration of foraging on 'key resources' at the base of catenas during the dry season (Scoones, 1989b; Jarman and Sincair, 1974). Resources within habitat patches, such as termite mounds or forage under trees, are also used selectively. Spatial heterogeneity thus has an impact on the interactions between population and resources, at different levels in a hierarchy of scales. The persistence of cattle populations can be interpreted in terms of the exploitation of environmental heterogeneity at different spatial scales.

There are two components to local strategies for averting mortalities and sustaining cattle populations: regional and local adaptive movement.

Persistence of cattle populations is thus assured by the expansion of the spatial range of resource use (through migration) and through the exploitation of local spatial heterogeneity and 'key resource' patches (through local foraging behaviour).

The effects of the major two-year drought of 1982-4 were ameliorated for some herds through movement. Data on herd survival rates showed that movement early in the drought period increased the chances of cattle survival. Three strategies were contrasted: those that moved animals before November 1982, those that moved animals after this time, and those that did not move animals at all. Herd owners who were able to exploit the opportunities of movement to better grazing were able to reduce the impact of stress year mortality. Cattle survival was 40% with maximum herd movement, 23% with late movement, and only 3.3% with no movement (Scoones, 1993: 392).

The drought period of 1987 was studied in detail through observation of cattle foraging behaviour in four herds. Cattle were followed for one day each month and the habitat types used at regular time intervals were observed (Scoones, 1990). The results show a distinct pattern of seasonal use, guided by herding during the cropping season and free grazing at other times of year. Table 4.5 shows the preference in terms of time spent feeding in relation to

Table 4.5: Preference indices for each habitat and study herd for three seasons[a]

Herd	*Season*[b]	*Upland*	*River dambo*	*Drainage*	*Contour*	*Home site*	*Field*
Clay 1	C	0.51	4.09	6.72	0.00	0.77	0.00
	ED	0.00	0.00	3.97	12.24	0.47	1.41
	LD	0.15	2.14	6.29	7.92	1.65	0.00
Clay 2	C	1.34	5.43	0.33	1.04	0.00	0.18
	ED	0.38	0.00	0.09	9.49	0.56	1.36
	LD	0.73	0.17	0.58	7.12	0.00	1.11
Edge	C	1.09	7.41	2.29	0.48	1.67	0.00
	ED	0.31	0.00	1.53	6.69	0.96	1.03
	LD	1.10	12.90	0.54	2.67	3.41	0.10
Sand	C	0.54	10.31		1.47	5.05	0.00
	ED	0.10	0.69		21.10	0.00	4.87
	LD	0.39	6.53		18.03	3.61	0.73

Notes: [a]Preference index = % time spent feeding in a particular habitat patch in that season as a proportion of total availability of that patch within 2km radius of home kraal. If PI > 1 habitat patch is preferred, if PI < 1 it is avoided and if PI = 1 it is used in proportion to its availability.

[b]C = Cropping season (herded); ED = Early dry season (free grazing); DS = Dry season (free grazing).

different habitat types for four herds (two from a clay veld area, one from a sand veld area and one from an intermediate, edge zone).

The two clay veld herds show a major preference for the riverine grazing habitat during the rainy season. One clay veld herd returns to this during the dry season. The sand veld herd uses the *vlei* (valley bottomlands) resource in the same way, as key grazing in the early rains and in the dry season. Drainage line grazing is also important for one of the clay veld herds throughout the year. All herds show a marked preference for arable land grazing during the dry season, with a switch from grazing field stover following the harvest to contour ridge grazing as the dry season progresses. No herd shows a large preference for topland grazing at any time.

Two patterns of exploitation are observed. One strategy relies on particularly valuable 'key resources', such as riverine strips, *vleis* or drainage lines, to provide fodder during a particular seasonal niche. This is shown especially by the clay veld (1) and the sand veld herd. Another strategy involves a more flexible response to spatial variation, with foraging spread over the range of habitat types and regular switching between patches observed. This is shown by the clay veld (2) and the edge zone herd.

Cattle survival, particularly at the key seasonal times at the end of the dry season and the early rains, is thus heavily influenced by the availability of fodder in 'key resources' (and access to such patches) and the ability to move flexibly between different habitat patches. The ability of the land to support animals (i.e. components of 'carrying capacity') is critically dependent on these factors.

The reasons why there are so many animals in the communal areas is thus linked to cattle's ability to exploit resource patches within a heterogeneous landscape in a flexible, opportunistic way, offsetting both the impact of density dependent regulatory factors and episodic density independent stresses that act to reduce population size.

Summary of conclusions

Cattle populations in southern Zimbabwe seem to have been rarely at equilibrium. The impact of regular droughts, disease or destocking has regularly reduced populations. A number of conclusions about the dynamics of cattle populations can be drawn:

- Populations may be regulated by density dependent factors acting on birth and death rates, but tend to be limited by density independent factors related to climatic variability.
- Regulatory factors are difficult to detect in systems where patchiness, environmental variability and complex dynamics interact.
- Potential ecological carrying capacity, according to simple models, is 2-3 times higher than official recommended carrying capacity.
- Official recommended levels relate to 'economic carrying capacity' for commercial beef production and are therefore irrelevant for communal area situations.

- Cattle populations are able to persist at high levels because of the availability of 'key resource' patches that provide forage at particular times of year and because of the ability to move adaptively between different portions of the landscape.

If this interpretation of the dynamics of the cattle populations of southern Zimbabwe is accepted, this has a number of implications for management and policy. Firstly, official recommendations for 'carrying capacity' require revision. A fixed figure is probably irrelevant, as populations are rarely at 'equilibrium' and fluctuate dramatically according to climatic (and other) factors.

Secondly, an adaptive approach to management is a more appropriate response to the dynamic variability of savanna systems. An event-oriented approach focuses on management adjustments to particular impacts (e.g. drought 'stress years') at particular times. Management with an emphasis on opportunistic and flexible responses is thus likely to be more effective than establishing a fixed strategy (Walker et al., 1986; Westoby et al., 1989).

Lastly, flexible responses to extend the spatial scale of use (in drought) or to switch between different habitat patches need to be supported by policy measures that protect tenure rights and access, and that facilitate movement during periods of stress. Similarly, measures to ensure access to 'key resource' patches within pastoral landscapes are essential. Measures to ensure system sustainability should be focused on the protection of the productivity in such key resource areas.

5
The Concept of Rangeland Carrying Capacity in Sub-Saharan Africa - Myth or Reality *

Peder N. de Leeuw and John C. Tothill

Introduction

Ever since colonial administrators and Western-trained scientists became involved in sub-Saharan Africa in the early twentieth century and were faced with the task of governing countries where livestock production was a major economic enterprise, the proper utilisation of rangelands became a major concern. While during most of the colonial era devastating epidemics (like rinderpest and pleuropneumonia) kept the growth of livestock populations in check, during the 1950s and 1960s regional campaigns of eradicating these major cattle diseases created a continuous increase in livestock numbers. It is estimated that between 1963 and 1970 the annual growth of cattle populations averaged 2.8%, but slowed down to 1.25% per annum from 1970 to 1980 (Antenneh, 1984).

While up to the 1950s constraints of feed shortages for livestock were outweighed by those of disease, concern over looming imbalances between supply and demand of grazing resources led to large investments in projects of water development, many of which were financed by foreign aid. This provision of water to hitherto under-used grazing lands further stimulated herd growth, but at the same time encouraged more settled modes of production and reduced mobility. While watering rights of man-made traditional wells were tightly enforced and served as a control on over-exploitation, water sources that were publicly financed were open to all, breaking the age-old equilibrium between water and rangeland use (Sandford, 1983b; Swift, 1984).

The need for establishing criteria and ways for determining the carrying capacity of African rangelands became more strongly felt when in the 1960s and 1970s several regions were hit by droughts (1960-61, 1973-76 in East Africa; 1969-73 and 1979 in West Africa) causing enormous stock losses.

This chapter investigates common approaches to determine carrying capacity, elucidates the problems associated with applying the concept in sub-Saharan Africa, and suggests ways of broadening the concept to improve its applicability in the continent.

Definition and assumptions

Carrying or grazing capacity (CC) is defined as the maximum possible stocking of herbivores that rangeland can support on a sustainable basis (FAO, 1988). Estimates of CC are commonly based on the assumption that livestock

---

* This chapter is an amended version of de Leeuw and Tothill 1990.

require a daily dry matter (DM) intake equivalent to 2.5% to 3.0% of their bodyweight. Thus, for a tropical livestock unit (TLU) of 250 kg of weight, 2.3 to 2.7 t of dry feed per annum is needed. To calculate an appropriate balance between forage supply and demand three multipliers are additionally required to adjust for:

(i) grazing efficiency (the proportion of total herbage livestock can harvest)

(ii) forage loss (due to trampling, fouling, decomposition, etc.)

(iii) proper use, which is the maximum proportion of forage that can be grazed without causing rangeland deterioration (FAO, 1988).

Although each of these three factors needs consideration, most estimates have used a single multiplier that combines adjustments for all. For example, Le Houérou and Hoste (1977) assumed that, in the Sahel, total dry matter (TDM) contained 40% edible forage, while in Southern Ethiopia, Cossins and Upton (1987) stated that one TLU required 8 t DM per annum, which converts to a utilisation rate of about 30%. Van Wijngaarden (1985) proposed a proper use factor of 45% of TDM during the dry season, based on his finding that when use was higher, perennial grass cover declined in the subsequent growing season.

In the arid zone of Niger, where wind erosion and loss of soil organic matter are prime concerns, protective cover equivalent to 0.2 t DM ha^{-1} should be left at the end of the dry season. A natural decomposition rate of 4% per dry month was allowed, which over 9 dry months converts to a 30% decline of the biomass at the end of the wet season (Wylie et al., 1988). Hiernaux (1982) in Mali described a similar method of adjustment for rangelands with mainly annual species. Variable monthly rates of disappearance were assigned, resulting in a 50% decline in standing biomass from October to March. Thereafter, during the hot and humid period of the late dry season, rates increased so that at the end of June only 10% of the original biomass was left.

As a result of these adjustments and corrections, estimates of CC do vary. For instance, in the Sahel, along the 400 mm isohyet with an average TDM of 1.1 t ha^{-1} (Table 5.1), three authors (Le Houérou and Hoste, 1977; Wylie et al., 1988; Hiernaux, 1982) predicted a year-round CC of 5.2, 4.4 and 7.0 ha TLU^{-1}, (19, 23, 14 TLU km^{-2}) respectively, which compares reasonably well with 5.5 ha TLU^{-1} (18 TLU km^{-1}) for East Africa given by Pratt and Gwynne (1977) for the same annual rainfall.

Average carrying capacity using TDM

To estimate CC, empirical relationships between annual or seasonal rainfall and total end-of-season herbaceous TDM ha^{-1} are commonly used (Table 5.1). Prediction for three areas in Africa indicates that over a rainfall range of 200 to 800 mm, TDM in t ha^{-1} in Kenya is higher than in West Africa due to higher fertility and greater water-holding capacity (WC) of the soils and lower evaporation rates because of altitude; conditions that give rise to a

Table 5.1: Estimation of total DM production from annual or seasonal rainfall (t DM ha^{-1})

Rainfall (mm)	*200*	*400*	*600*	*800*	*Source*
West Africa	0.6	1.1	1.7	2.2	1
Zimbabwe					
(WC:100mm)[a]	0.5	1.7	2.2	2.5	2
(WC:200mm)	0.7	2.6	3.2	3.7	2
Kenya	1.1	2.3	3.6	-	3

Note: [a] WC denotes water holding capacity
Sources: 1. Le Houérou and Hoste (1977); 2. Dye and Spear (1982); 3. de Leeuw and Nyambaka (1989)

predominantly perennial grass cover. However, TDM production in West Africa differs less if it is further adjusted for soil water-holding capacity. Blair Rains and Kassam (1979) suggested that TDM values for West Africa be adjusted 50% upwards for deep heavy upland soil and 50% downwards for shallow stony soils; thus for 400 mm of rain, TDM would range from 1.7 to 0.5 t DM ha^{-1}. This is a wider range than that given for Zimbabwe by Dye and Spear (1982), where TDM yields on heavy soils was about 50% greater than on sandy soils (Table 5.1). Similarly, van Wijngaarden (1985) in Eastern Kenya predicted for a seasonal rainfall of 200 mm, a yield of 1.1 t DM on deep sandy clay soils and 0.6 t on shallow gravelly soils.

Further corrections are required for the effects of run-off and for woody cover. Run-off is usually accounted for by substituting effective for total rainfall. For instance, in the 300 to 400 mm zone in Burkina Faso, Grouzis and Sicot (1983) showed that run-off was 10% to 23% of total rainfall and the results of their regression equation of effective rainfall on TMD is similar to that of Le Houérou and Hoste (1977). For a detailed account of the effects of run-off and run-on on primary productivity in the Sahel, reference is made to Penning de Vries and Djitèye (1982) and to Hiernaux (1982).

Reduction of herbaceous TDM due to woody cover has been discussed by several authors. In Kenya, van Wijngaarden (1985) found that for each 10% increase in woody cover, perennial herbaceous cover declined by 7%, reaching zero cover when the woody canopy reached 90%. Blair Rains and Kassam (1979) developed a non-linear relationship where at 20% woody cover TDM was reduced to 73%, and to a mere 17% when the ligneous stratum reached 60%. Timberlake and Reddy (1986) used this relationship to adjust their estimates of CC for grazing lands of Mozambique. An even larger effect of the tree strata on herbaceous TDM was reported in the sub-humid zone in Mali with a mean rainfall of 1,100 mm. Without tree cover, recorded yields were 5 t DM ha^{-1}, which dropped linearly to 1 t when tree cover increased to 40% (Penning de Vries and Djitèye, 1982). Rutherford (1978) confirmed the inverse relations between herbaceous production and woody cover in Southern Africa, while for his carrying capacity map in Botswana, Field (1978) assumed a reduction of 5% in CC per every 100 units of trees/shrubs ha^{-1}. However, decline of herbaceous biomass due to woody cover is offset somewhat by increased browse production, the magnitude of which depends on species composition and degree of woody cover.

The livestock-oriented approach

Using total herbaceous forage productivity as the single criterion to predict the livestock support capacity has been criticised because biomass quality and feeding value for livestock are largely ignored. Thus, if the concept of CC were to be expanded to include these qualitative values, then the maximum allowable stocking rate could be set at a level reflecting the specific production objectives of livestock managers. CC should be linked to nutritive value, in particular to the crude protein (CP), energy and mineral content of the TDM (FAO 1988).

Predictions of forage quality in the Sahelian environment were based on the finding that, below 300 mm of rainfall per season, soil moisture was the most important factor controlling plant growth, while above this isohyet, soil nitrogen and to a lesser extent phosphorus became major contributing factors (Penning de Vries and Djitèye, 1982). Plant N uptake was associated with herbage N-content at the end of the growing season, while the soil N-balance could be calculated from expected gains and losses. N-uptake by herbaceous plants is mainly governed by rainfall and over a rainfall from 400 to 1,000 mm, N-uptake increases from around 10 to 20 kg N ha^{-1}. As a consequence, for any given seasonal rainfall, N-content of herbage at the end of the growing season becomes a function of total yield. Due to this N-dilution, Penning de Vries and Djitèye (1982) showed that across this rainfall gradient in West Africa TDM increased from 1.1 to 4 t ha^{-1}, while the end of season N-content declined from 1.0 % to 0.5 % (6.2% to 3.1% CP).

The high quality of grass biomass in arid areas was confirmed in Kenya. In Turkana, average protein in cattle diets consisting almost entirely of grasses was 11.6% CP (1.8% N), and declined somewhat in the long dry season but remained above 8% CP or 1.3% N (Coppock et al., 1986). Similar levels of 7 - 8.8 % CP (1.1 - 1.4% N) in dry season grass were reported by van Wijngaarden (1985) working in arid Eastern Kenya. These trends in herbage quality along rainfall gradients explained the growth performance of cattle. In Northern Niger at 320 mm rainfall, steers gained up 80 kg during the short rainy season and lost little weight during the dry season, provided sufficient forage was available (Klein, 1981; Wylie et al., 1983). In contrast, cattle grazing rangelands receiving 400 to 500 mm further south, gained 60 to 90 kg during the rains, but net annual gains per head were much lower due to weight losses of 30 to 50 kg during the dry season (P. Hiernaux, unpublished data).

Traditional livestock owners in the semi-arid and sub-humid zones in West Africa have adopted several strategies to reduce this quality constraint in the dry season; these strategies include herd mobility or relying on feed sources other than dry standing grass. Transhumance patterns involve long-distance

Table 5.2: Fodder composition of cattle diets in two agropastoral systems in three seasons in semi-arid Nigeria (in % of grazing time)

Annual rainfall	*700 mm*			*1,000 mm*		
Season	*wet*	*early dry*	*late dry*	*wet*	*early dry*	*late dry*
Months	*Jun -Sep*	*Oct -Jan*	*Feb -May*	*Jun -Sep*	*Oct -Jan*	*Feb -May*
Upland savanna	76	24	59	96	56	22
Regrowth after fire	6	1	11	-	-	-
Valley grassland	8	28	3	1	5	58
Browse	10	6	12	3	4	5
Sub total	100	59	85	100	65	85
Cereal residues	-	31	9	-	23	5
Legume residues	-	4	4	-	10	-
Cotton residues	-	6	2	-	2	10
Sub total	-	41	15	-	35	15

Source: Adapted from Van Raay and de Leeuw (1974)

treks to better feed sources in floodplains such as the inner Niger delta in central Mali, the Benue river and the shores of Lake Chad. In Northern Nigeria sedentary stock owners achieved better dry season diets for their stock by herding them to cultivated land to consume crop residues after harvest in the early dry season and by switching to browse, valley grassland and regrowth after fire later in the dry season (Table 5.2). Similarly in central Mali, rice straw and grass regrowth in fields that were irrigated during the wet season were the main sources of dry season feed for cattle, while small stock were taken to upland grazing grounds to enable them to browse (Table 5.3). The presence of these extra feed resources accounted for the relatively high stocking rates of 10 ha TLU^{-1} (10 TLU km^{-2}) in the dry season and 6.4 ha TLU^{-1} (16 TLU km^{-2}) in the wet season in this area (Bourn and Wint 1985a).

Sources of variation

A shortcoming of the carrying capacity concept is that it is often regarded as a static assessment, while it should be seen as varying in space and over time. Spatial variability occurs at several levels and is caused by differences in soil type, grazing pressure and incidence of fire, resulting in variable species composition and degree of plant cover. In Sahelian rangeland, where annual

Table 5.3: Annual diet composition (in %) of cattle, sheep and goats in a secondary production system in semi-arid Mali

Fodder source	*Cattle*	*Sheep*	*Goats*
Herbaceous grazing	53	59	11
Browse	4	34	87
Millet residues	6	7	2
Rice residues and grass regrowth	37	-	-

Source: Adapted from Dicko-Touré (1980)

species with either C_3 or C_4 photo-synthetic pathways and contrasting germination strategies occur together, small-scale spatial and temporal variation in yields is large (Breman et al., 1984). For instance, Hiernaux (1983) demonstrated that in a Sahelian dune landscape, total TDM measured over a 3-year period was 1.8 to 2.0 t on the crest, 0.5 to 2.0 t along the slope and 2 to 7 t ha^{-1} in inter-dune depressions that benefited from run-on moisture.

In rangelands with perennial grasses, the degree of plant cover is often an important source of variation. Van Wijngaarden (1985) found that apart from differences due to the depth, texture and fertility of soils, the response to rainfall was mainly governed by the amount of perennial plant cover. In pastoral areas, reduction of plant cover is often unavoidable and related to 'sacrifice areas' along livestock trek routes, around water points and homesteads.

Variability on a large scale can sometimes be deduced from detailed vegetation maps. In a limited area (260 km^2) in Burkina Faso, rangeland yields were recorded and mapped in 1977 and 1980 when seasonal rainfall was 440 and 310 mm respectively. The distribution of yield classes showed a range from 0.2 to 5 t ha^{-1} and differed markedly between years (Table 5.4).

In recent years the variability of TDM over large areas has been assessed through AVHRR/NOAA imagery[1] using regression equations of the Normalised Difference Vegetation Index (NDVI) of TDM which predicted herbage production in steps of 0.4 t ha^{-1} (Tucker et al.,1983). In Niger, in a rainfall zone of 150 to 400mm, distribution patterns in 1985 and 1987 indicated that in about 40% of the pastoral area surveyed, TDM was 0.8 ha^{-1} or above (Table 5.5). However, in 1987 there was little plant growth north of 15.5° latitude, whereas in 1985 there was evidence of primary production as far north as 18° corresponding to similar trends in adjacent Sahelian countries (Tucker et al., 1986).

1. Advanced Very High Resolution Radiometer of the National Oceanic and Atmospheric Administration.

Table 5.4: Spatial distribution of herbaceous biomass in Northwest Burkina Faso in 1977 and 1980

Year	*1977*		*1980*
Mean rainfall (mm)	440		310
t DM ha^{-1}		*(in % of area)*	
0.2 - 0.5	3		22
0.5 - 1.0	29		30
1.0 - 1.5	35		31
> 1.5	33		17
Weighted mean	1.1		0.8

Source: Adapted from Figure 1 in Grouzis and Sicot (1983)

These inter-annual variations are caused by many factors, the major one being the effect of rainfall. In the Sahel, the Coefficients of Variation along the 200 to 600 mm gradient are usually 20% to 30% and for a mean of 400 mm of rainfall there is a 10% probability that the amount is less than 300 mm or more than 560 mm. Hence at P > 0.8, for the mean of 1.1 t as given in Table 5.1, TDM can vary between 0.9 and 1.5 t ha^{-1} (Hiernaux, 1982; Penning de Vries and Djitèye, 1982). Such inter-annual variations in yields were confirmed by Grouzis and Sicot (1983) in semi-arid Burkina Faso. For a mean rainfall of 400 mm, the median biomass was 1 t DM ha^{-1}, ranging from 0.8 to 1.4 t at 80% probability or corresponding to CCs of 7.4 and 4.1 ha TLU^{-1} (13.5 and 24.4 TLU km^{-2}) (Table 5.6). In 1977 the dry season stocking rate was estimated at 3.9 ha TLU^{-1}, which implies that the area was likely to be overstocked during most years.

Table 5.5: Distribution of standing herbaceous biomass in Northern Niger in 1985 and 1987 derived from AVHRR/NOAA satellite data[a]

t DM ha^{-1}	*1985*[1]		*1987*[2]
		(% of area)	
0 - 0.2	3		34
0.2 - 0.8	57		30
0.8 - 1.4	38		24
> 1.4	2		12
Weighted mean	0.7		0.5

Note: [a]Location: 14.5^{0} - 16.0^{0}; 5^{0}E
Sources: 1. Wagenaar and de Ridder (1986), 2. Wylie et al. (1988)

Table 5.6: Probability of rainfall, biomass and carrying capacity in Northern Burkina Faso

Probability %	*90%*	*50%*	*10%*
Rainfall mm	295	410	575
Biomass (t DM ha^{-1})	0.77	1.02	1.38
CC, ha TLU^{-1}	6.7	5.6	4.1

Source: Grouzis and Sicot (1983)

In semi-arid areas with a bimodal rainfall distribution, variability is much greater. For instance, in Eastern Kenya an annual rainfall of 580 mm recorded over 73 years showed a CV of 52% due to the combined variability of two rainy seasons of 310 and 270 mm with CVs of 65% and 74% respectively. From a regression equation of seasonal rainfall on TDM, it was calculated that yields ranged from 0.3 to 7.5 t ha. In 25% of the years, annual yields were below 2 t and in 22% above 4 t, with a mean of 3.1 t ha^{-1} (Table 5.7). As a result, the mean CC would be about 2 ha TLU^{-1} which was close to actual stocking rates measured in Maasai group ranches with a similar rainfall (de Leeuw et al., 1984; Bell, 1982). However, given this high inter-annual variability, it implies that in one year out of four, these ranches run the risk of being seriously overstocked, if the livestock population remains static and is not affected by drought or adjusted by sales.

Actual stocking rates

The question arises to what extent these theoretical estimates of CC conform to observed stocking rates. In Niger, the average CC over 1985 to 1987 was

Table 5.7: Frequency distribution of annual herbaceous DM yield in semi-arid Eastern Kenya

Yield class	*% of years*	*t DM ha^{-1}*	
t DM ha^{-1}		mean	(SD)
< 1.0	4	0.4	(0.3)
1.0 - 1.9	20	1.6	(0.3)
2.0 - 2.9	29	1.6	(0.3)
3.0 - 3.9	25	3.6	(0.3)
4.0 - 4.9	12	4.4	(0.2)
> 4.9	10	6.4	(0.8)
Mean (n = 73)		3.1	(1.5)

Source: Adapted from de Leeuw and Nyambaka (1988)

calculated to be 6.7 ha TLU^{-1} (15 TLU km^{-2}) in an area with a rainfall from 150 to 400 mm along a North-South gradient and producing an average TDM output of 0.7 to 0.8 t DM ha^{-1} (Wylie et al., 1988).

No data on the livestock population were available for 1985-7, but in 1981-2 (prior to the serious drought in 1984) the actual stocking was estimated at about 13.5 ha TLU^{-1} (7.5 TLU km^{-2}) or close to half the theoretical capacity. These estimates came from systematic reconnaissance flights (SRF) during which livestock were counted in an area of 80,000 km^2. It was established that 60% of the total mass consisted of cattle, while camels, smallstock and donkeys accounted for 13%, 24% and 3% respectively (de Leeuw and de Haan, 1983; Milligan, 1982).

Similar surveys were carried out in the Gourma region in Mali with 150 to 500 mm of rainfall (Bourn and Wint, 1985b). Average wet and dry season stocking rates in 1983 to 1984 were estimated as 18 and 25 ha TLU^{-1} (5.6 and 4.0 TLU km^{-2}) respectively, and cattle and small stock accounted for 76% and 21% of the total livestock mass. Closer scrutiny of these SRF data showed that in the wet season the actual stocking increased steeply along the North to South gradient from 62 ha TLU^{-1} (1.6 TLU km^{-2}) in the 150 to 200 mm zone to 11 ha TLU^{-1} (9.1 TLU km^{-2}) in the 400 to 500 mm belt.

Much higher stocking rates were recorded in Northeast Senegal where development of artesian boreholes has been going on for more than two decades (Sutter, 1987). In an area of 30,000 km^2 along a rainfall gradient of 200 to 500 mm, the average stocking rates was 7.4 ha TLU^{-1} (13.5 TLU km^{-2}) in 1983. However, aerial surveys indicated an uneven distribution across the area. In about 40% of the area stocking was less than 20 ha TLU^{-1} (5 TLU km^{-2}), whereas 25% of the area carried 15 to 50 TLU km^{-2} and another 5% carried over 50 TLU km^{-2} (Sharman, 1983). During the favourable years of 1980-82, average TDM was in the order of 1.0 to 1.3 t ha^{-1} and well distributed over the area. For instance in 1981, only 8% of the area yielded less than 0.4 t ha, while close to 40% produced 1.2 t ha^{-1} or more (Tucker et al., 1983). For this period, CC was calculated at about 5 ha TLU^{-1} (20 TLU km^{-2}) or well above the actual stocking rate. However, in 1983 rainfall was extremely low (a mean of about 100 mm for 22 rainfall stations) with the result that in almost 80% of the area, TDM was below 0.2 t ha^{-1}, causing serious shortages of feed, a high rate of mortality and an exodus of stock to areas further south (Vanpraet et al., 1983). Simulation of TDM output in this area failed to predict such low yields. Cornet (1984), using a sophisticated water balance model, calculated that TDM yields of below 0.6 ha^{-1} would occur in 1 year out of 20, whereas in 4 years out of 5, yields exceeded 1.1 t ha^{-1}.

Thus, in general, Sahelian rangelands were utilised below their capacity in most years, but were heavily overstocked in exceptional drought years. However, there is no way of avoiding such opportunistic strategies of pastoral people, since the alternative of conservative stocking adapted to probabilistic incidence of almost complete rainfall failure would result in a reduction in the average human support capacity of those rangelands (Sandford, 1983a).

Higher stocking rates of herbivores with increasing annual rainfall was also shown by Bell (1982) using data for Eastern and Southern Africa derived from

Coe et al., (1976). These trends were confirmed for all sub-Saharan countries by Bourn (1976) up to 1,200 mm, above which densities declined, supposedly due to increased tsetse challenge. However, when the sites used by Bell (1982) were separated into those with high and low soil fertility, the linkage between annual rainfall and level of stocking was less strong. In the rainfall range of 700 to 900 mm densities of herbivores (including wildlife) were equivalent to 25 to 33 TLU km^{-2} on soils with high nutrient status developed over volcanic, rift and alluvial parent materials as compared to 5 to 8 TLU km^{-2} for low fertility sites over basement complex receiving 800 to 1,100 mm of rainfall. Bell (1982) concluded that TDM production alone is a poor indicator of edible feed for herbivores and that nutrient quality therefore is an important determinant of carrying capacity. This interaction between rainfall and soil nutrient status reinforces the relationships outlined by Penning de Vries and Djitèye (1982) as discussed above.

Although in line with Bourn (1976) the risk of trypanosomiasis has usually been linked to the low stocking rates in the sub-humid zone in West Africa, the low quality of feed resources is another explanatory factor, as most of this zone is characterised by poor soils derived from basement complex. However, with increasing population density the tsetse challenge is diminishing, particularly in Nigeria, and stocking rates in the sub-humid zone have been rising rapidly over the last two decades. Nevertheless, during 1980-84, cattle densities varied from 4 to 40 head km^{-2} with an overall mean of only 10 head km^{-2} for the entire zone (Blench et al., 1985). Since the average TDM is about 3 t ha^{-1}, a CC based on biomass alone would be in the order of 2 ha TLU^{-1} or about 65 cattle km^{-2}. However, extensive and repeated SRFs in Nigeria have indicated that cattle distributions are governed by many factors, chief among which are the density of dry season watering points, the rural population density and the infrastructure (ibid.).

Discussion and conclusions

Two major approaches, either plant- or animal-oriented, to determine CC have been described. In the first, feed resources are assessed and permissible use by livestock is pre-conditioned by proper use factors that are geared to prevent over-exploitation and to safeguard sustained range productivity. Feed output can be estimated from empirical or simulation models using rainfall or soil-moisture balance techniques, enabling longer-term forecasting based on past climatic records. Livestock use is predicted from known daily feed intake requirements and proper use is mostly set at 30-50% of the available herbaceous biomass, the rate of use depending on the known or assumed fragility of the ecosystem.

The animal-oriented approach follows a similar pathway in that the permissible amount of feed taken off by livestock is determined as in the 'plant-oriented' approach. However, since permissible stocking rates are supposed to ensure the adequate nutrition of stock to attain desired production goals (in terms of survival, milk, growth, etc.) herbage quality and nutrients output also need consideration. It is evident that, in view of the seasonal variations in nutrient content of natural herbage, animal-oriented permissible stocking rates on the whole tend to be lower than those predicted by the plant-oriented approach. For instance, if sustained milk yields during long dry seasons are a desired goal, large areas of land may be required to provide lactating cows with a diet of adequate nutritional value.

Both approaches pose problems which are associated with characteristics of the African environment and the production systems found in the various ecological zones. These problems relate to scale (the area of assessment), species mix, mobility, land tenure and production goals of actual producers.

The concept of CC assumes that livestock are kept within fixed areas of land with recognised boundaries. Such conditions rarely prevail in Africa. Mobility of stock together with communal land tenure and fluid rights of access to grazing and water do not facilitate the computation of meaningful carrying capacities. In both transhumant and agro-pastoral systems, stock rely on several niches within the ecosystem, while in areas with cropping, residues from crops are an important source of feed. The contributions of these different feed sources to the annual diet are difficult to quantify.

The scale of the CC assessment has an important bearing on its validity and usefulness. If direct interrelationships between animal output and feed supplies are aimed at, resources within the grazing orbit of the individual producer need to be determined. This is more easily achieved for individual mixed farmers than for nomadic or transhumant pastoralists. For these latter groups, communal use of grazing resources implies that only aggregate values of carrying capacity are meaningful, the size of the area being dependent on the boundaries of common use.

While, for example, mixed farmers in semi-arid Kenya have legal title to their land enabling an assessment of feed resources of their stock, group ranches contain 30 to 50 families using an area of 10,000 ha or more with loosely defined grazing rights within their boundaries. In most of Southern Africa, stock-owning farmers have stable usufructuary rights to the land they cultivate, while communal grazing lands are shared with many owners. This implies that the scale at which carrying capacity must be estimated is that of the entire village area. In West Africa, stock-owning farmers entrust their stock to pastoralists either on a permanent basis or during the crop-growing season when livestock may do damage to crops. Consequently, a calculation of CC has to include different land units sometimes many kilometres apart.

In developed countries, multi-species exploitation of rangelands is rather uncommon. Even if more than one species of livestock are kept, they are often allocated separate areas of grazing land. In Africa, multi-species enterprises are the rule. In the more arid zones, camels and smallstock predominate, while in better watered areas, cattle and smallstock are kept by smallholder farmers and by Maasai in group ranches alike. Consequently, several livestock enterprises are combined within the same management unit. As feed preferences between species differ, assessing the fraction of edible feed becomes more complicated, in particular where browse from woody species is an important component. In the developed world, the concept of CC has been applied mostly to cattle production enterprises and as a consequence, graminoid herbage chosen by cattle has become the norm for identifying desirable species. Herbs, dwarf shrubs and taller woody species are classed as undesirable and would decrease CC in proportion to their abundance. In multi-species enterprises, emphasis on graminoid herbage would underestimate CC if the grazing value of herbs and woody species is negatively assessed.

Nonetheless, the concept of carrying capacity is useful for planning purposes, calculations of the average productivity of land in terms of feed resources, and expected output of livestock. While long-term averages may be required for general forward projections, the limitations of such data should be fully recognised in the light of sharp fluctuations over time and space. Upper and lower limits of CC need to be considered and linked with probability of expected levels of herbage production together with the interactive effects of over- and under-utilisation in the medium term.

Finally, the ultimate validity of the carrying capacity concept rests on the recognition that feed resources are governed by an interlinked set of environmental factors, which determine the upper and lower limits of primary as well as secondary productivity in the short term but even more so in a long-term perspective. In view of the decline of per caput food output in Africa, there is a risk that policies to enhance the human support capacity of African grazing land in the short term will ignore the long-term consequences. Therefore, although the concept of carrying capacity may be of limited utility for African conditions, the underlying principles on which it is based need full acceptance, without which no sustained resource management can be accomplished.

6
An Examination of the Carrying Capacity Concept*

Gerrit B. Bartels, Brien E. Norton and Gregory K. Perrier

Introduction

Carrying capacity is one of those terms in range management that has the appearance of objectivity but in practice means different things to different people, depending on whether their primary interest is livestock production, resource conservation, or wildlife management. Even among livestock producers, it may mean one thing to a commercial rancher in America and quite another to a subsistence pastoralist in sub-Saharan Africa. Some people use it to quantify the exploitation of forage in the current year or grazing season; others focus on the long-term level of utilisation at which the impact of grazing on the forage resource is acceptable. There is also inconsistency in the methods employed to estimate carrying capacity. Furthermore, for a given methodology, the procedure may produce a reliable estimation in one ecological zone and a much less credible figure in a region of greater annual variability in forage production. Some have argued (e.g., Stoddart, 1960) that for practical purposes it cannot be estimated at all.

Despite conflicting interpretations and uses, carrying capacity has evolved from a simple index of the limit to the exploitation of grazing land by grazing animals and become imbued with economic significance, a critical parameter in resource assessment and planning, particularly in Africa. Given the importance of carrying capacity as a tool to define and implement livestock development projects, it seemed worthwhile to review its historical origins, theoretical basis, and suitability in African range management.

This chapter begins with a note on the derivation of 'carrying capacity' and then examines the concept through a comparison of published definitions, followed by a commentary on methods used to estimate it. We then discuss whether carrying capacity is an appropriate and applicable parameter for range managers working in sub-Saharan Africa, assuming that it can be presented without confusion and estimated with reasonable accuracy.

Origins

The debate over the prospects of feeding an expanding human population in Europe goes back to at least the seventeenth and eighteenth centuries. Forty years after Malthus' 1798 landmark treatise on exponential growth, Verhulst

---

* Portions of this paper have previously appeared, with some rewording, in Bartels et al. 1990.

(1838) proposed a logistic equation whose asymptote is probably the earliest reference to an upper limit of population size based on fixed resources (Hutchinson, 1978). When Pearl and Reed rediscovered this model in 1920 they defined the asymptote 'K' as the maximum number of individuals that can be supported on a fixed area. According to Dhondt (1989), Odum (1953) was the first ecologist to equate K with 'carrying capacity,' a term used since 1906 (or earlier) by range managers who were probably unaware of the theoretical foundation initially developed by Verhulst.[1] Ecological texts (e.g., Colinvaux, 1986; Ricklefs, 1979) have followed Odum's lead and defined carrying capacity as the maximum number of individuals that can persist under certain specified conditions, particularly limited space and food supply. This definition implies that carrying capacity is the total resources available divided by the minimum maintenance requirement of each individual. In contrast, the maximum sustainable yield occurs when the growth rate of a population is maximal under the conditions prevailing, and is found at a population size that equals K/2, i.e., half the carrying capacity predicted by the logistic model (Wilson, 1980).

Under strictly controlled conditions with a constant food supply, the logistic growth model gives a reasonable fit in the cases of microbial organisms and relatively uncomplicated small animals such as certain invertebrates (Hutchinson, 1978; Colinvaux, 1986). The carrying-capacity population (K) becomes a stable equilibrium point and the food supply is the major density-dependent factor.

An important reason why the logistic model works so well when tested in laboratory experiments is that the food is replenished; there is no negative feed-back of population size on the availability of food. This is a very different situation from that normally encountered when populations of large herbivores are grazing rangeland, where animal numbers not only respond to the availability of resources but have an impact on them as well. The simple principles embedded in the logistic growth model are not so easily extrapolated to natural ecosystems, and the adoption of carrying capacity as a broad ecological concept has generated an assortment of definitions without a consensus on its exact meaning.

Confusion surrounding the definition of carrying capacity

At first glance one may be surprised to find that the concept of carrying capacity is the subject of contention among range managers. The idea seems to be relatively straightforward. In a very general sense, 'carrying capacity' is the population that can be sustained by the resources which the animals use within a particular area. However, adding climatic variability and a resource-

---

1. In the 1906 *Yearbook of the U.S. Department of Agriculture*, p. 226, Cotton noted that overgrazing by livestock or 'the ravages of prairie fires' may reduce the carrying capacity of the range. Sampson (1913) talked about decline in carrying capacity in the context of overgrazing 'during the "free-for-all" grazing period of earlier days.'

feedback interaction to the application of carrying capacity in an ecological context creates a complexity shaped in part by the tolerance of the specific herbivores to deteriorating resources and the tolerance of the particular ecosystem to increasing grazing pressure. The temporal dimension contributes another ingredient to the definition, especially if carrying capacity is viewed as a long-term parameter that must accommodate climatic uncertainty, instead of being a variable that changes on an annual or seasonal basis. The specific objectives of the resource manager add another component; for example, a manager whose goal is to maintain a range resource in good condition will perceive a lower carrying capacity than a manager who wants to keep large numbers of animals.

The confusion produced by all these facets of the concept is evident when one compares the twelve definitions of carrying capacity extracted from the grazing management literature and listed in Appendix A. They range from the vague '...the optimum stocking level to achieve specific objectives given specified management options' (Scarnecchia, 1990), to the specific

> ...the user-specified quality biomass of a particular species, under the influence of social or behavioral constraints, for which a particular area, having user-specified objectives, will supply all energetic and physiological requirements over a long (but specified) period (Giles, 1978),

and include the cryptic '...the stocking rate at the optimum grazing pressure' (Mott, 1960).

Several reasons may be offered for the existence of this state of confusion. One problem is the dichotomy between livestock and wildlife management (Caughley, 1976; Macnab, 1985; Dhondt, 1989). As a management concept, carrying capacity was introduced by range managers at the beginning of this century (Cotton, 1906) and applied to domestic animals in North America, predominantly within the context of commercial beef production. In the 1930s it was adopted by wildlife management, a discipline with objectives that are sometimes completely different from those in livestock-oriented range management. In commercial ranching the primary objective is to maximise net income. In wildlife management, on the other hand, harvesting high-priced animals is seldom the sole objective; other goals include the control of animal numbers, conservation of species, and habitat maintenance or improvement. Moreover, wildlife do not receive the regular care and supervision of domesticated animals, a distinction that has led to definitions such as that by Edwards and Fowle (1955):

> [Carrying capacity is] the maximum number of grazing animals of a given class that can in a given ecosystem *survive* through the least favorable environmental conditions occurring within a stated time interval. For practical purposes this time interval is usually one year. (Emphasis added)[2]

---

2. The 1964 glossary of the Society for Range Management presents a similar definition: ...'the greatest number of animals that can survive the greatest period of stress each year...'

Secondly, much confusion arose when ecologists in the second half of this century began to equate the term carrying capacity, already an established concept in range management, with the very different equilibrium value K of the logistic growth model. In a strict ecological sense (i.e., as the asymptote K), the carrying capacity is the number of individuals beyond which no major increase in population size can occur without an improvement in habitat or resources. Animals in such a population are able to survive long enough to reproduce but they are not necessarily vigorous or in good health -- an undesirable status for a herd of livestock on a commercial ranch. When range managers in the Western United States refer to carrying capacity they are usually talking about the maximum sustainable yield point; individual animals are in good condition and highly productive. Population size at this managed level is usually between 50 and 75% of that found at the resource-restricted carrying capacity (Odum, 1983; Macnab, 1985). Thus from the point of view of theoretical ecology, carrying capacity as employed in range management is a misnomer; conversely, range managers, whose scientific discipline is based on ecological principles and whose goal is to optimise animal off-take and sustain range productivity, consider a carrying capacity determined by the point of starvation as impractical.

The importance of making a clear distinction between the two 'carrying capacities' has been emphasised by, among others, Dasmann (1981) who named one 'optimum density,' which he equated with carrying capacity in livestock-oriented range management and which has also been called 'economic carrying capacity' (Caughley, 1979),[3] and the other 'subsistence density.' The latter has also been called 'environmental carrying capacity' (Clark, 1976), 'ecological carrying capacity' (Caughley, 1979), 'maximum carrying capacity' and 'potential carrying capacity' (Riney, 1982). While the subsistence density represents the population size at which an equilibrium between food demand and supply is reached without external interference (i.e., K of the logistic model), the economic carrying capacity or optimum density denotes an equilibrium held in place by human intervention. Numerous managed equilibria are possible, depending on management goals and the levels of animal harvesting prescribed by those goals, but they are all characterised by lower numbers of animals and higher levels of accessible forage biomass than are found at the subsistence density (Caughley, 1979, Macnab, 1985).

The commercial livestock bias of Western-educated range managers is a

---

3. Although Dasmann does not take commercial interests into his reckoning of the optimum, most economists would prefer to use the phrase 'optimum carrying capacity' only when there is an economic basis for its determination (e.g., Workman, 1986). For Workman, the point of optimum economic yield is at a stocking rate below that which produces maximum biological yield per unit area, the latter being theoretically K/2. Caughley (1979) used the term "economic carrying capacity" referring to stocking rates providing maximum sustained yield rather than maximizing financial return. In range management literature the optimum carrying capacity is generally assumed to be the stocking rate at the point of maximum sustainable yield, unless otherwise specified, and in this paper we follow this convention.

stumbling-block when they try to assess appropriate stocking levels for production systems that operate much closer to the environmental carrying capacity, which is the case of traditional pastoralism in sub-Saharan Africa. Vegetation and soil parameters used to define good range condition for populations kept near the maximum sustainable yield point are unnecessarily restrictive when the goal is to keep the maximum number of animals. However, the potential for resource degradation is greatly enhanced when the stocking rate is increased from the economic carrying capacity[4] to that of the environmental carrying capacity. Whether this risk will ever be expressed depends to a large extent on weather sequences, the intrinsic capacity of the forage resource to withstand stress, and the relative importance of density-independent factors in regulating animal population size, as discussed below.

Another source of confusion in the concept of carrying capacity is the question of its stability. An unfortunate consequence for natural resource managers of equating carrying capacity with K in the logistic model of population growth, is that it implies the mathematical stability of an asymptote, or at least an oscillation of narrow amplitude around a stable point. As pointed out earlier, natural plant-herbivore systems do not comply with two assumptions underlying the logistic growth model: (i) the food supply is not constant, partly because animals affect the availability and composition of their forage; and (ii) the physical environment varies, often stochastically. In its closest resemblance to K, the ecological carrying capacity is, at best, an average point around which actual numbers of herbivores fluctuate within certain boundaries. This may be the case in regions characterised by little climatic variation between years and vegetation tolerant of heavy grazing pressure, as may obtain in parts of Europe. In extreme environments, however, such as semi-arid regions where droughts are normal, plant growth is highly variable and so are the mortality and birth rates in animal populations. In such environments, density-independent (climatic) factors, by the imposition of sudden shifts in the ecological limits to population size, are likely to play a greater role in regulating herbivores than density-dependent ones. It is extremely doubtful that the determination of a carrying capacity in such environments has any relevance for management of the resource. It has been suggested that the lowest equilibrium point, e.g. in drought years, should be used as a measure of the resource's carrying capacity. In highly variable environments like the Sahel, however, such an approach is wasteful of forage and certainly unacceptable to livestock producers. Use of an 'average' carrying capacity is likely to lead to unsatisfactory results as well because, as Stoddart (1952) pointed out, 'there is no reason to believe that overuse in one year can be compensated by underuse in another year.'

---

4. The phrase 'economic carrying capacity' is used here strictly in relation to Western livestock production that emphasises high productivity per animal. We recognize that in the economy of traditional pastoralism in Africa, the livestock herds are treated as personal assets equivalent to bank accounts, and therefore the goal of maintaining large numbers of animals is a valid economic objective. Within these alternative terms of reference, the 'economic carrying capacity' is in fact close to the ecological carrying capacity.

Different perspectives on the time element can also cause difficulties in understanding carrying capacity. The relevance of this point is best demonstrated by a question posed by Macnab (1985): What does the ecological carrying capacity really represent? Is it a short-term condition in which animal numbers reflect a resource that is still in good or fair condition, or does it represent the situation in which an equilibrium is reached after a long period of decline in range condition?

Finally, some range managers treat carrying capacity and grazing capacity as synonyms (e.g., Stoddart et al., 1975; Pratt and Gwynne, 1977; Holechek et al., 1989), while others make a distinction between the two terms, although for completely different reasons (e.g., see Heady, 1975 compared to Vallentine, 1990). As may be seen in Appendices A and B, the American Society for Range Management has changed its mind twice on this issue over the past thirty years. The Society's 1964 glossary states explicitly that grazing capacity and carrying capacity are not synonymous, but the next edition of the Society's glossary equates the two, while the 1989 edition distinguishes them again (Society for Range Management, 1964, 1974, 1989). It is remarkable that the range management profession has not yet reached a consensus on defining such a basic concept. For example, whereas the Society for Range Management (1989) includes 'harvested roughages and concentrates' in its definition of grazing capacity, Vallentine's recent textbook (1990) places 'harvested forages and other feedstuffs' within his definition of carrying capacity.

Elements of a useful definition

Based on the foregoing discussion, we conclude that a definition of carrying capacity which is to be used as a guideline in natural resource management must be both specific and complete.

Our present review examines general definitions (Appendix A) which try to convey the concept with, in our opinion, varying degrees of success. We propose that a good general definition should also fulfil the role of a model or template into which the details for a particular management or ecological situation are inserted. Vague definitions such as those of Mott (1960) and Scarnecchia (1990) are not helpful in understanding the carrying capacity concept nor are they useful as references for a particular case of range resource management. A general definition should encompass the elements of animal production goal, acceptable resource condition, and time.

Because of the relation between animal numbers and animal production, a definition of carrying capacity which does not include a production objective is rather meaningless (see, for instance, Hodgson, 1979). The same reservations were expressed for an African context by the 1983 International Workshop on Land Evaluation for Extensive Grazing (Butterworth and de Ridder, 1983). Unless the definition relates specifically to subsistence density, reference should always be made to some measure of animal performance. Using this criterion, only the definitions given by Sampson (1923), Dasmann (1945), Vallentine (1990, in terms of grazing capacity), and, possibly, Heady (1975) are clearly referring to a carrying capacity other than the ecological variant. The way in which the definitions by the Society for Range Management (1964, 1974, 1989) are phrased merely adds to the confusion since they appear to allude to subsistence density (see Note 2) rather than the optimum density, although a stocking rate close to the latter is probably intended.

The ecological consequences of large herbivore populations grazing the forage resource need to be considered in a definition, and acceptable conditions of the resource stated. In the illustrative list of definitions in Appendix A, range condition is either not mentioned (e.g., Edwards and Fowle, 1955; Society for Range Management, 1964; Pratt and Gwynne, 1977) or it is defined rather vaguely (e.g., Hadwen and Palmer, 1922; Sampson, 1923; Dasmann, 1945; Society for Range Management, 1974, 1989). Vallentine (1990) accommodates acceptable resource condition in his definition of grazing capacity but not when defining carrying capacity.

Inclusion of a time frame is important. We believe that in the context of range management, definitions of carrying capacity are clearly deficient without the proviso that future forage production is not to be adversely affected by present stocking rates (e.g., Cowlishaw, 1969[5]). A time reference will clarify whether the carrying capacity figure is being used as an equivalent to the annual or seasonal stocking rate based on available forage alone, i.e., applying only to a short-term situation, or whether the effects of defoliation on the dynamics of plant growth and the feedback on the herbivore population have been taken into account, as they must be in determining the long-term productive capacity of the resource. In the definitions presented in the Appendices, a time period is either not specified (e.g., Mott, 1960; Scarnecchia, 1990) or varies from short - a year or less (e.g., Sampson, 1923; Edwards and Fowle, 1955; Cowlishaw 1969; Society for Range Management 1964, 1974, 1989) - to long (e.g., Giles, 1979; Dasmann, 1945; Vallentine, 1990, definition of grazing capacity). Hadwen and Palmer (1922) and Pratt and Gwynne (1977) are rather ambiguous on this issue. Caughley's (1979) idea of carrying capacity is that of an equilibrium point reached by natural feedback interactions between herbivore and the resource, which presumes an unspecified time period to achieve this balance.

Although we have reservations with regard to the applicability of the carrying capacity concept in range management, especially as discussed below in relation to sub-Saharan Africa, we feel that resource management would be better served by more accurate and more specific definitions. We believe that Giles' (1978) general definition provides an excellent blueprint to follow.

Can carrying capacity be adequately estimated?

Despite the confusion over conflicting definitions of carrying capacity, one finds that it is routinely estimated in sub-Saharan Africa, based on measures or estimates of annual or seasonal plant biomass production and daily dry matter intake. These estimates are then adjusted by a 'proper use' factor,

---

5. Although Cowlishaw did not make this point explicitly in his definition, it emerges from subsequent discussion in his paper.

which is the level of utilisation recommended on the basis of sound resource and livestock management. But how accurate are such estimates?

In a generally ignored essay, Stoddart (1960) listed a total of fourteen factors that need to be considered in determining carrying capacity, including topography, distribution of water and livestock, effects of wild herbivores, season of use, grazing management practices, and non-livestock demands on the range resource (e.g., tourism in America and fuelwood gathering in Africa). Pratt and Gwynne (1977) and Purnell (1983) stressed the importance of water point distribution in determining carrying capacity in East Africa. In their papers discussing the concept of carrying capacity for pastoral systems in sub-Saharan Africa, Blair Rains (1983) and de Leeuw and Tothill (1990) addressed major shortcomings in estimation methods. They noted the difficulty of determining available forage per animal owing to high annual and spatial variability in plant production, seasonal changes in forage quantity and quality, livestock species mix, and the use of supplemental feeds. Breman et al. (1983) pointed out that species composition and forage quality aspects are often neglected in calculations of carrying capacity; they present an example where these omissions resulted in a gross overestimation of carrying capacity. Another problem is that forage available for consumption is virtually always based on estimates of peak plant biomass. However, when a range is used at high grazing pressures (yet within the long-term sustainable stocking rate) forage production is usually less than the peak estimated from exclosures or lightly grazed areas.

Probably even more important is the realisation that the choice of the proper use factor, which in most cases is an educated guess at best, can have a profound effect on the estimate of carrying capacity. For instance, the decision to apply a proper use of 45% instead of 30% increases the calculated carrying capacity by half. The 'proper use' is not some constant value for a plant species or rangeland reflecting its intrinsic tolerance to defoliation, but rather a parameter that will vary according to growing conditions, range condition or species composition, season of defoliation, and type of herbivore (Cook and Stoddart, 1953, Pratt and Gwynne, 1977).

As Sampson observed in the first textbook on range management (1923)

> ...in view of the numerous variable and uncontrollable factors that present themselves nearly every season, it is not likely that the grazing capacity factor will ever be worked out to a high degree of scientific accuracy.

We strongly concur with Stoddart (1960) when he claimed that

> ...there is no method whereby any technician can go into a new country and measure anything which will automatically give him the grazing capacity.

A similar conclusion was reached by Smith (1984) in a review of the methods used by United States land management agencies to determine the carrying capacity of rangelands.

Nevertheless, numerous experts have made attempts in sub-Saharan Africa to estimate carrying capacity on the assumption that, if stocked at carrying capacity, pastoral production systems could achieve the dual goals of range conservation and a relatively high and constant off-take. On the whole, these efforts have demonstrated that carrying capacity cannot be estimated accurately enough for planning purposes.

Measurements of the same resources have yielded divergent estimates for carrying capacity, while no confidence intervals are ever given. De Leeuw and Tothill (1990) report estimates of carrying capacity for similar environments in the Sahel of 14, 19 and 23 Tropical Livestock Units per km^2. (One TLU is 250 kg of equivalent livestock weight). In planning range development for a 100 km^2 area, that represents a difference of 900 TLUs between the highest and lowest projections. This kind of disparity could be a controversial issue if development plans call upon pastoral households to adjust their herd size to one particular, and probably conservative, carrying capacity value.

Theoretically impossible claims of overgrazing levels have been made, e.g., stocking rates at 160% greater (USAID, 1980) and 200% greater than carrying capacity (Field, 1980[6]). Several authors (Riewe, 1961; Hart, 1972; Jones and Sandland, 1974) have proposed a quadratic relationship between stocking rate and livestock production per unit area. This implies that the optimum stocking rate, expressed as the stocking rate at which maximum gain per hectare is achieved, is half the stocking rate at which animals start to lose weight. Or, in other words, at a stocking rate twice as high as optimum carrying capacity (maximum yield per area), animal weight gain is zero. The assumptions of this model may be valid under conditions where the forage on offer does not vary significantly in quantity and quality. In situations where high stocking rates cause negative feedback (reducing forage productivity) the difference between the stocking rate of maximum yield per unit area and the stocking rate at which there is no weight gain, may be even smaller (Roberts, 1980). These models suggest that stocking rates greater than twice the optimum carrying capacity are biologically unsustainable, and therefore such claims should be viewed with suspicion.

Is carrying capacity appropriate to sub-Saharan Africa?

The validity of carrying capacity as a management concept is based on the premise that range productivity is dependent upon stocking rate in a causal relationship; heavy grazing promotes degradation and reducing the stocking

---

6. Field (1980) made careful calculations of actual stocking rate and carrying capacity (in terms of Standard Stock Units of 450 kg) for each Region in Somalia, based on estimates of populations of camels, cattle, sheep and goats, their age-class distributions and estimated liveweights by age-class; the proportion of diet for each livestock type coming from herbaceous and browse resources; the dry matter feed requirements of each livestock type; estimated production of herbaceous and browse material; and a proper use factor of 50%. He concluded that actual stocking rates were more than twice the carrying capacity for 13 of the 16 Regions, averaging three times the carrying capacity overall. At the extreme, Bari Region SR was estimated at eight times its carrying capacity.

rate leads to range improvement. In sub-Saharan Africa this premise can be challenged on two counts. First, the great variability in rainfall, especially the incidence of drought, can overshadow the influence of herbivory on the range resource. Thus range productivity becomes more a function of climate (a density-independent factor) than of stocking rate (Ellis and Swift, 1988, Westoby et al., 1989), and the climatic and stocking rate effects interact and exert episodic rather than continuous impacts. Livestock populations are more likely to be depressed by drought than by overgrazing, and when a series of wet years follows the drought the regenerative capacity of the herds is often too low (Wilson, 1986) to keep up with the recovery of forage production. It is not surprising therefore that some studies suggest that livestock populations in Africa are below estimated carrying capacity (e.g., Coughenour et al., 1985).

The second challenge is that as forage resources decline under increasing grazing pressure, pastoralists move their animals to more favourable areas. In essence, pastoralists apply their own pragmatic version of 'carrying capacity' and withdraw their herds from the threat of inadequate forage. This practice is the basis of a rational drought-survival strategy. We do not deny that degradation due to overgrazing occurs in sub-Saharan Africa, but suggest that it occurs only under special rather than general circumstances, namely, when animals are concentrated so that the forage on a specific area of rangeland rather than other feed resources or drinking water is the limiting factor; i.e., when the mobility of the herds is constrained.

Many pasture production systems in sub-Saharan Africa use crop residues, fallow farmland, miscellaneous patches (roadsides, interspaces between fields, etc.), and 'special resources' (Scoones 1989b) such as seasonal wetlands, as well as rangelands. The rangelands are not necessarily the limiting resource for livestock production (e.g., Scoones, 1989b, van Raay and de Leeuw, 1970), except perhaps during a severe drought. If the overall stocking rate is restricted by resources other than rangeland, would not a value for carrying capacity be irrelevant if based on estimated range forage biomass alone?

Can carrying capacity be meaningfully applied?

The concepts of carrying capacity and stocking rate, as generally understood in Western range management, assume that a unique population of livestock is directly associated with a defined grazing area for a specific period of time. In systems where livestock are herded rather than fenced, and land tenure is communal or open access, the grazing areas used by the livestock of a specific household can vary greatly between years, as can the livestock demand placed on a given grazing site. Households can adjust the area of land used to achieve a balance between livestock nutritional needs and the energy demands incurred by the search for forage. Households can also adjust the time spent grazing a seasonal forage resource. Even when one household holds its grazing management constant, variability in use by other households, both resident and transient, can greatly alter the livestock demand placed on a given grazing area.

Reciprocal agreements for resource use are a further complication. In many pastoral societies, one group will allow another group, in need, access to grazing or watering resources. It is understood that when the group offering this hospitality is itself in need, the other group will reciprocate by allowing access to its forage and water resources. Reciprocal agreements allow pastoralists to survive in environments of high annual variability in the amount and distribution of precipitation. These agreements, along with communal use, make it difficult to apply the concept of carrying capacity to pastoral systems. Social customs and cultural values can place an occasional and unpredictable high livestock demand on a grazing resource regardless of agreed stocking rate guidelines, preempting the ecological wisdom of subscribing to a particular carrying capacity.

Application of the carrying capacity concept on North American rangelands assumes that one decision-maker or a cohesive decision-making body has sole control of the grazing resource. In systems where land tenure is communal or open access, numerous households, each making more or less independent management decisions, are using the same grazing resource. When there is one manager of the grazing area, this manager tends to view land as the limiting resource, and the balance between livestock and land becomes a critical factor. When there are numerous decision-makers sharing a grazing resource, each decision-maker has available an area of land greater than his/her livestock can use. From the viewpoint of one household, land is in surplus and thus has a low value. Livestock and labour become the main constraints on household welfare, and the balance between livestock and labour becomes the critical management issue. In such situations, households strongly resist adjusting livestock to the land base, an irrational practice viewed at the household level, and therefore it is very difficult to get compliance with efforts to reduce livestock to a calculated carrying capacity.

Another issue which brings into question the effectiveness of the carrying capacity concept in pastoral systems is the investment role of livestock. In Western ranching systems, livestock generally offer a low return on capital (Workman, 1986) and any excess capital created through livestock production flows into more lucrative investments. In most of rural Africa, on the other hand, a livestock enterprise can offer the greatest return on investment. For this reason, capital flows into livestock rather than out of it. It is irrational to expect a pastoral household to convert capital growth from livestock into the local currency, unless this cash can flow into investments more productive than livestock. Lacking more lucrative investments, each household attempts to increase its herd size (wealth) with indifference to the restrictions that a carrying capacity limit may impose.

Though there have been numerous attempts, we know of no case in which a government agency in Africa has successfully persuaded pastoral households, or a pastoral group, to voluntarily reduce livestock numbers on rangeland to satisfy an estimated carrying capacity.

Conclusion

There is no consensus on the meaning of carrying capacity in the range resource management literature, but the dominant approach focuses on management of the stocking rate to maintain high animal production per unit area and protect the range resource from overgrazing. The 'optimum carrying capacity' is a stocking rate which satisfies the interests of commercial ranching; the higher 'ecological carrying capacity' is the stocking rate determined by the limitations of available forage, and is more applicable to traditional pastoralism.

The suggestion is made that a general definition of carrying capacity has two functions. The first is to convey the concept; the second is to serve as a blueprint for more specific definitions to be used as management guidelines. Such definitions should encompass the three elements: time, animal production goal and acceptable resource condition.

We have concluded that carrying capacity, as conceived in Western range management, is of questionable validity in livestock production systems in Africa, that it is virtually impossible to estimate it accurately, and that the concept cannot be meaningfully applied in pastoral systems, except perhaps in special circumstances where animals are confined to rangeland, their mobility is restricted and there is relatively little variation in forage production. The enormous expense devoted to estimating carrying capacity in sub-Saharan Africa has contributed little to livestock development and has diverted resources from other priorities. Let us admit the problems with the carrying capacity concept in sub-Saharan Africa, and stop trying to apply it.

We still need to determine how to ensure that permanent degradation to forage resources does not occur because of occasional excessive stocking rates. We recommend taking an opportunistic approach, basing the annual grazing strategy on that year's forage production. This allows households to adjust to spatial variability in forage production, provides the most balanced distribution of livestock to forage, and supports more people than a conservative approach based on the estimated or drought-year carrying capacity for rangeland.

In order to minimise resource degradation we recommend Stoddart's (1960) suggestion of starting with an initial stocking rate (the extant livestock population) and monitoring changes in livestock production and the forage resource. Livestock could be monitored for long-term trends in weight gain, mortality of young animals and milk production. The forage resource could be monitored for long-term trends in species composition and percentage cover of ungrazed vegetation. Long-term trends for the soil properties of infiltration capacity and erosion should also be monitored. Permanent monitoring programs are needed to develop databases adequate to distinguish long-term trends from annual variability. Our pessimism in setting carrying capacity was shared by the 1983 African workshop on land evaluation for extensive grazing. It also recommended monitoring range condition and livestock production, because carrying capacity estimates are always approximate and because carrying capacity changes over time (Butterworth and de Ridder, 1983).

When monitoring detects declines, induced by overgrazing, in long-term livestock or forage productivity, or in infiltration capacity, or increases in soil erosion, then measures must be taken to conserve the forage resources upon which households depend. Such system degradation occurs where livestock are concentrated to the point that plant biomass becomes the factor limiting livestock production. Concentrations of livestock in African pastoral systems tend to be very localised and are often the result of government programmes, i.e., water development, grazing reserves, etc. All government programmes in pastoral areas should be scrutinised for their effects on livestock densities, and concentrations that make plant biomass on rangelands limiting should be avoided. When such concentrations do occur, it might be easier to assist some households to leave a degrading area than to persuade all households to reduce their herd size. Efforts to avoid degrading concentrations of livestock might focus on managing household density, rather than controlling livestock density.

Appendix A: Some selected definitions of carrying capacity

Hadwen and Palmer (1922: 29)
'... the number of stock which a range will support for a definite period of grazing without injury to the range.'

Sampson (1923: 328)
'... the number of stock of one or more classes which an area will support in good condition during the time that the forage is palatable and accessible, without decreasing the forage production in subsequent seasons.'

Dasmann (1945: 400)
'... the maximum number of grazing animals of a given class that can be maintained in good flesh year after year on a grazing unit without injury to the range forage growing stock or to the basic soil resource.'

Edwards and Fowle (1955: 597)
'... the maximum number of animals of given species and quality that can in a given ecosystem survive through the least favourable environmental conditions occurring within a stated time interval. For practical purposes this time interval is usually one year.

Mott (1960: 607)
'The stocking rate at the optimum grazing pressure.'

Cowlishaw (1969: 207)
'... the number of animals of a specified type that can subsist on a unit area and produce at a required rate over a specified period, usually a season, a year, or longer.'

Giles (1978: 195)
'... the user-specified quality biomass of a particular species, under the influence of social or behavioral constraints, for which a particular area, having user-specified objectives, will supply all energetic and physiological requirements over a long (but specified) period.'

Caughley (1979: 5)
'... the point where the rate of production of forage equals the rate at which that forage is consumed.'

Society for Range Management (1964)
'In its true sense, the maximum number of individual animals that can survive

the greatest period of stress each year on a given land area. It does not refer to sustained production. In range management, the term has become erroneously synonymous with grazing capacity.'

Society for Range Management (1974, 1989)

'The maximum stocking rate possible which is consistent with maintaining or improving vegetation or related resources. It may vary from year to year on the same area due to fluctuating forage production.'

Scarnecchia (1990: 554)

'... the optimum stocking level to achieve specific objectives given specified management options.'

Vallentine (1990: 458)

'... the total amount of [consumable forage] produced and available for grazing from ... a specified land area, [including] harvested forages and even other feedstuffs used in conjunction with grazing, thereby providing a means of summarizing total ranch or animal enterprise capacity.'

Appendix B: Some definitions of grazing capacity

Mott (1960)

'Grazing capacity is sometimes a synonym for carrying capacity'.

Society for Range Management (1964)

'The maximum stocking rate possible without inducing damage to vegetation or related resources.'

[This definition was adopted by the International Workshop on Land Evaluation for Extensive Grazing, Addis Ababa, 1983, which called it carrying capacity, not grazing capacity (Siderius, 1983: 5).]

Society for Range Management (1974)

'Grazing capacity = carrying capacity.'

Stoddart, Smith, and Box (1975: 183)

'... the maximum animal numbers which can graze each year on a given area of range, for a specific number of days, without inducing a downward trend in forage production, forage quality, or soil.'

Heady (1975: 115)

'... the number of animals that produces the greatest return without damage to physical resources and in concert with other values received from the land.'

Pratt and Gwynne (1977: 276)

'Grazing capacity (or livestock carrying capacity [sic]): the number of stock of a given class or classes, expressed in livestock units or head, which a rangeland unit will support for the period of grazing (or feeding) allowed.'

Society for Range Management (1989)

'.... the total number of animals which may be sustained on a given area based on total forage resources available, including harvested roughages and concentrates.'

Vallentine (1990: 462)

'1. The total number of AUMs[7] produced and available for grazing from a pasture unit, grazing allotment, the total ranch, or other specified land area.

2. The maximum stocking rate possible without inducing damage to the soil, vegetation or related resources or deleteriously affecting grazing animal response. Must consider factors such as annual fluctuations in forage production, kind or mix of animal species, season and system of grazing, and grazing distribution to be accurate.

Synonymous with carrying capacity, at least in part.'

---

7. An Animal Unit Month (AUM) is the amount of forage consumed per month by a cow weighing 454 kg (1,000 lb), or equivalent weight of other type of livestock.

7
A New Approach to Evaluating Rangeland Productivity in Sahelian Countries

Nico de Ridder and Henk Breman

Introduction

Agriculture in the Sahelian countries has considerably changed during the last decades as a result of several droughts, over-population and slow economic development. With the focus of livestock activity moving southwards from the Sahelian zone to the savannas, animal husbandry drastically changed. The nomadic and semi-nomadic peoples lost substantial parts of their herds as a result of high mortality rates, as well as by forced sales caused by worsening terms of trade between livestock and arable products. However, the number of livestock kept by arable farmers in the south has increased considerably, producing draught power and manure to improve stagnating or reduced crop yields. Some wealthier citizens, the new livestock owners, still regard livestock as a worthwhile investment, but have other production objectives than the former owners, the nomadic and semi-nomadic peoples. Others have even started up intensified milk production systems around the big cities. Thus, livestock change hands and occupy new territories, while livestock owners adopt other production objectives.

In addition to these changes, a cause of stress and dispute for pastoralists since colonial times has been the discrepancy between official laws and the local, traditional pastoral codes with regard to grazing rights and water use. Opportunities for optimum pasture use are reduced by these discrepancies, while advancing pasture degradation now urgently calls for legislation and control.

This description of a rapidly changing world (Breman et al., 1990) emphasises the need for better knowledge of pasture and livestock production in West Africa, a region renowned for its excellent livestock producers (Breman and de Wit, 1983). Governments, national services, development programmes and donors all want to know what role livestock can play in rural development, how semi- and nomadic pastoralists can be helped and which bottle-necks need to be cleared.

The need for a pasture evaluation approach for West Africa

In general, livestock production is mainly determined by nutrition, rather than water availability or animal health problems (Breman and Traoré, 1986a, 1986b and 1987). Since extensive livestock systems depend mainly on fodder production from natural rangelands, animal production is primarily increased by the quantitative and qualitative improvement of pasture production. It is therefore necessary to have a method of evaluating both of these rangeland aspects.

Several handbooks have been written on rangeland evaluation. That by Boudet (1978) specially relates to the Savanna and Sahelian zone in West Africa. Although highly recommended as a first introduction to these natural rangelands, it is too purely descriptive to provide answers to the above-mentioned questions, in a quickly changing situation.

Other handbooks (e.g. Stoddart et al., 1975) focus on the pasture characteristics of ecological zones considerably different from those in the Sahelian countries. The particular climate of West Africa uniquely affects the pastures: the Sahelian and the Sudan zone is characterised by a short rainy season, with a rainfall gradient of 150 - 1200 mm yr^{-1} restricted to the summer, and high temperatures during the entire year. Such a climatic regime has a strong impact on pasture composition. In general, the predominance of annuals is assumed to be the result of overgrazing a natural perennial vegetation; it is also considered to be an intermediate stage in an evolution to a relatively more stable perennial grassland or forest, provided that stocking rates are decreased. However, the West African climate favours annuals more than perennials and this climatic effect is more pronounced with higher rainfall variability and sandy soils. Rainfall variability increases from the southern savanna (higher rainfall area) towards the northern Sahel (lower rainfall area), and sandy soils also increasingly prevail along this transect. Therefore, certainly in the Sahel, annuals are likely to predominate in the pastures, even when moderately or not grazed at all.

These differences underscore the need for a specialised handbook, which would enable evaluation of the productivity of West African pastures, taking into account the year-to-year fluctuations in their composition. Furthermore, it is necessary to classify the extent to which these 'annual pastures' are the result of natural forces or of overgrazing. Methods for estimating plant production, animal production and carrying capacity, as well as methods for identifying factors which limit production, need also to be applied in those areas where perennial grasses, shrubs and trees prevail.

Development of a pasture manual

The methods developed for a pasture handbook grew from a joint research project between Mali (Institut d'Economie Rurale) and the Netherlands (Centre for Agrobiological Research and the Agricultural University of Wageningen) from 1976 to 1980 (Penning de Vries and Djitèye, 1982). In the following years, there was additional research including field studies on the dynamics of annual pastures (Cissé, 1986) and desk studies (with simulation models), while the methods being developed were tested in the field (Breman and Traoré, 1986a, 1986b and 1987; Leloup and Traoré, 1989). This research and testing programme was concluded by the publication of a handbook in 1991 (Breman and de Ridder, 1991).

This chapter presents an overview of the handbook, demonstrating its particular structure with respect to the objectives (topics, user category etc.) and methodology, as illustrated in the evaluation of primary and secondary production and carrying capacity.

Structure of the Handbook

The handbook, intended to be an instrument for analysis and management, is designed to serve particular objectives and user groups. The objective is that, by using the handbook when analysing a region, it should be possible to estimate plant production and its quality, and the related animal production. Furthermore, it should be possible to identify factors limiting productivity, and ways of alleviating these limitations. Such an analysis becomes a management tool, as use of the handbook will enable conclusions to be drawn regarding the carrying capacity and the economic value of pastures, as well as enabling detection of changes in potential productivity.

The handbook is presented in four volumes, applying the evaluation methods at three different levels of precision, aimed at different user groups. The first volume reviews the specific character of the pastures in Sahelian countries, based on a broad analysis of the most important ecological zones. This volume is written for policy-makers involved in livestock and rural development.

The second volume is the handbook proper, written for technicians of local services and institutes, as well as for specialists within development projects and programmes. The user can evaluate pastures at three levels of precision. In the first evaluation level, only data on soil characteristics, climate and land use are considered, without the user undertaking any fieldwork. This broad method is useful in order for the user to obtain a preliminary impression of a region, with regard to potentials for and the problems with keeping livestock, as determined solely by the natural resources. In the second phase, the evaluation level is deepened by the addition of information gained from a year's field work. This level of precision should be used for regional rural development planning, project planning and for improving land use. Finally, in a third phase, the highest degree of evaluation precision can be reached on the basis of several years of fieldwork. This level of precision is needed to support research for regional rural development programmes and to monitor the pasture condition. Continuous monitoring is crucial to assess the effects of interventions designed to improve livestock production, or the effects of not intervening at all.

The scientific background of the evaluation methods used in the preceding volumes is elaborated in the third volume, which elucidates the choice of parameters. Since it is impossible for a handbook to anticipate and discuss all field situations which may exist, the information given in the third volume allows users to adjust the evaluation methods according to their own needs and situations, which enlarges the applicability of the handbook.

The final volume presents a synthesis of information, collected from the authors' experiments or deduced from the literature, on the characteristics of the most important herbaceous and woody species. Such information is required to apply the evaluation methods presented in volume two, and reference to this information may reduce the costs of chemical analyses.

Estimating primary production

Because of the unique character of the Sahelian and the Sudan zones, the evaluation method for estimating primary production in these zones differs from other methods in general use. The proposed method is initially based on landscape characteristics (geomorphology, relief, soil texture and depth) and climate (rainfall), rather than on vegetation characteristics (Fig. 7.1). Vegetation is subjected to dynamics of pasture production and species composition as well as limitations of water or nutrients, limitations which vary with time and place.

Both climate and landscape characteristics determine water availability: climate relates to rainfall and evapotranspiration, whereas landscape characteristics, through the run-off coefficient and storage capacity, determine the amount of water stored in the soil. For the Sahelian and the Sudan zone, 250 mm yr^{-1} of available water is considered the transition point between water-limited (< 250 mm yr^{-1}) and nitrogen-limited (≥ 250 mm yr^{-1}) production (Penning de Vries and Djitèye, 1982). For the two situations, different pathways and corresponding model equations are used to estimate nitrogen availability in the above ground biomass (N_b), nitrogen content and herb layer biomass.

Browse production and its nitrogen content is estimated on the basis of tree cover in relation to water availability. If competition between woody and herb species is expected, N_b of the herb layer is subsequently corrected. If perennial grasses account for an important part of pasture production, N_b is also corrected: the f-factor in the N_b-equation, representing the nitrogen losses, is then lower, since perennials redistribute nitrogen from the above-ground biomass to the roots at the end of the growing season.

In other more common evaluation methods, pasture productivity is determined on the basis of one year field observations of biomass, species composition (and sometimes also quality). The evaluation method for pastures in the Sahelian and the Sudan zone departs from the theoretically estimated N_b for years with average rainfall. The N_b is subsequently verified by field observations, if circumstances allow. Measured nitrogen (N_m) in the year of observation can be standardised to obtain values for an average year (N_s): methods are given in the handbook for standardising N_m for specific years, as influenced by the variable annual impact of species composition, grazing pressure, recent occurrence of bush fire, contribution of legumes and the total amount and distribution of rainfall, as observed during the fieldwork. After the standardisation procedure N_s is compared with N_b. If N_s differs from N_b, it is concluded that real soil fertility in terms of nitrogen availability is lower ($N_s < N_b$) or higher ($N_s > N_b$) than supposed, and consequently N_s is taken as the basis for estimating secondary production; otherwise, N_b is taken. In both situations, the nitrogen in the above-ground biomass of the herb layer, and consequently the biomass itself, can be estimated for dry and wet years with reference to production in an average year and also when stocking rates change.

Figure 7.1: Procedure to estimate biomass of herb layer (B) and browse, nitrogen in the above ground biomass (N_b) and nitrogen content

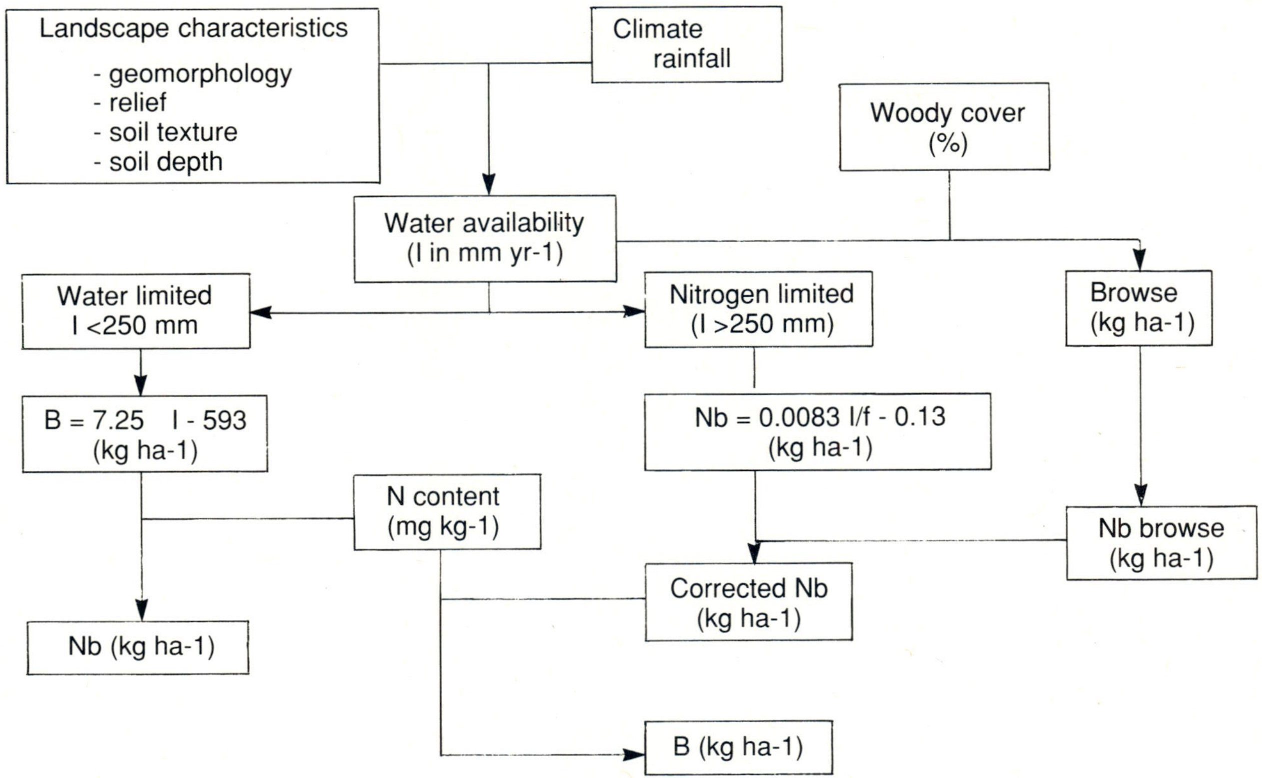

Without fieldwork, verification of the theoretical estimated N_b is rather difficult, since data published by others refer only to actual biomass production. However, with the estimated nitrogen content, a theoretical estimated biomass can be calculated and compared with field data published by others. This is done in the broad evaluation of the Sahelian and the Sudan zone, presented in volume one, which is treated here as an illustration.

Biomass of the herb layer in average rainfall years is calculated for five broad landscape units and four rainfall zones (200, 450, 750 and 1050 mm), which represent respectively the northern and southern Sahel and northern and southern Sudan. Each landscape unit's fraction of the total surface area of a rainfall zone enables the calculation of the average production per rainfall zone. Both the production per landscape unit and the average production per rainfall zone are presented in Fig. 7.2.

In this figure, data are shown from Boudet (1978) and Le Houérou and Hoste (1977). Boudet specifies production per landscape unit and climatic

Figure 7.2: Comparison between estimated biomass of the herb layer and data published by Boudet (1978) and Le Houérou and Hoste (1977)

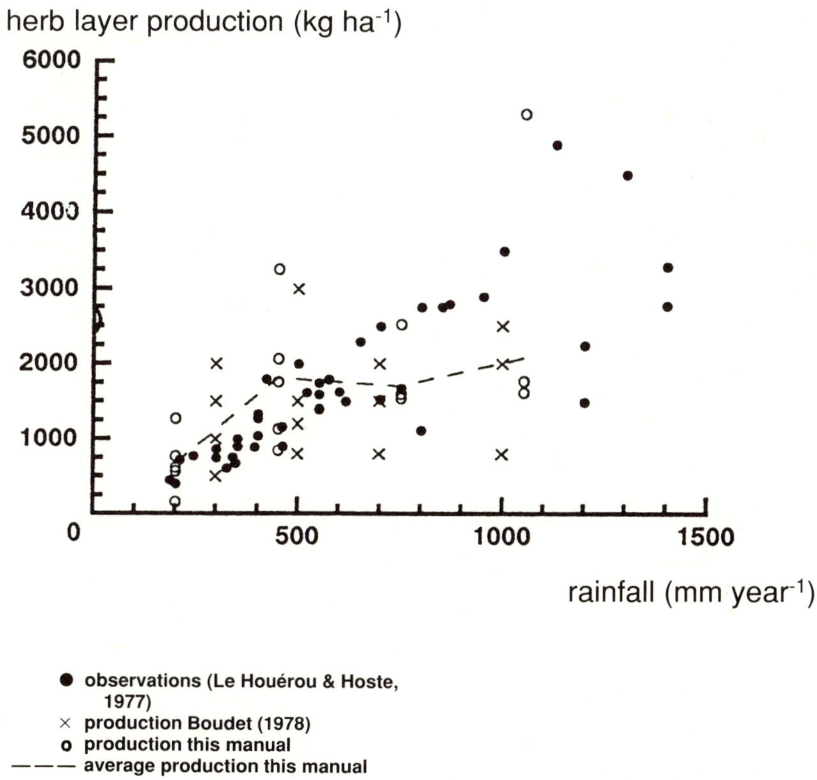

zone. Le Houérou and Hoste present a simple relation between observed rainfall and biomass production. The calculated values per zone and per landscape unit, based on this handbook, agree well with those observed by Boudet. Both sets of values remain within the limits published by Le Houérou and Hoste, with only one exception: production values for one landscape unit, the depressions, are systematically higher than the limit values presented by Le Houérou and Hoste, who excluded data obtained from depressions, in order to obtain a clear relationship between rainfall and production.

Estimation of secondary production

Biomass production and its nitrogen content (of herb layer as well as browse) can be used to estimate secondary production. For this purpose an intake model has been developed, which estimates intake on the basis of nitrogen content and digestibility of the selected forage. The step between forage on offer and selected forage will be considered later on.

Nitrogen content and digestibility as sole characteristics for fodder quality determine the upper limit of dry matter intake: whenever minerals (e.g. phosphorus) or vitamins are more limiting, real intake will be lower.

Since digestibility is rarely measured, an average relation between nitrogen content and digestibility is applied (Fig. 7.3a), as established for 225 samples. A similar approach was followed for browse samples, showing a generally lower digestibility with larger deviations (Fig. 7.3b). These general relations are used in the handbook examples, but whenever local field samples are available, the digestibility and nitrogen content of those samples is preferred.

Dry matter intake is converted to digestible dry matter and energy intake to assess weight gain or losses, milk production and herd growth.

Besides estimating secondary production on the basis of nitrogen content and digestibility of the forage selected, the reverse holds as well. When production goals are defined, required feed intake, nitrogen content and digestibility can be calculated for meeting these goals.

To estimate the amount of selected forage from the forage on offer, firstly the fraction of the total biomass which can be potentially consumed, of the herb layer on offer - the maximum consumable biomass - is calculated. Based on data from the literature, it is assumed that at most half of the biomass on offer can be consumed when pastures are grazed during the rainy season only, and one third when grazed during the dry season or year-round. This fraction represents the maximum total available biomass from which selection can be made.

The fraction of the maximum consumable biomass which will finally be consumed depends on the nitrogen content (and related digestibility) in the selected forage, as represented by the lines in Fig. 7.4.

When all maximum consumable biomass is consumed, the nitrogen content in the selected forage is equal to the average nitrogen content in the maximum consumable biomass. Nitrogen content in the selected forage is higher when, through selection, less of this biomass is consumed. In this way, intake per animal or tropical livestock unit (TLU), depending on the nitrogen content in

Figure 7.3: Digestibility (kg kg^{-1}) in relation to nitrogen content (g kg^{-1}) for herb layer species (a) and for leaves of woody species (b)

(a)

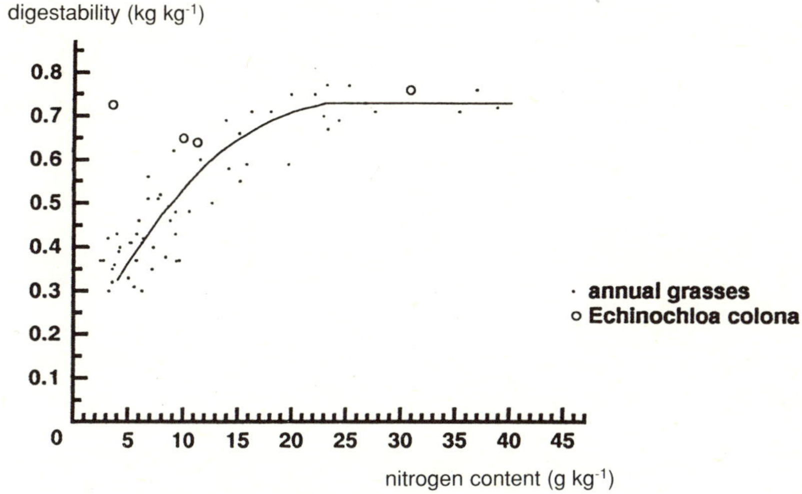

(b)

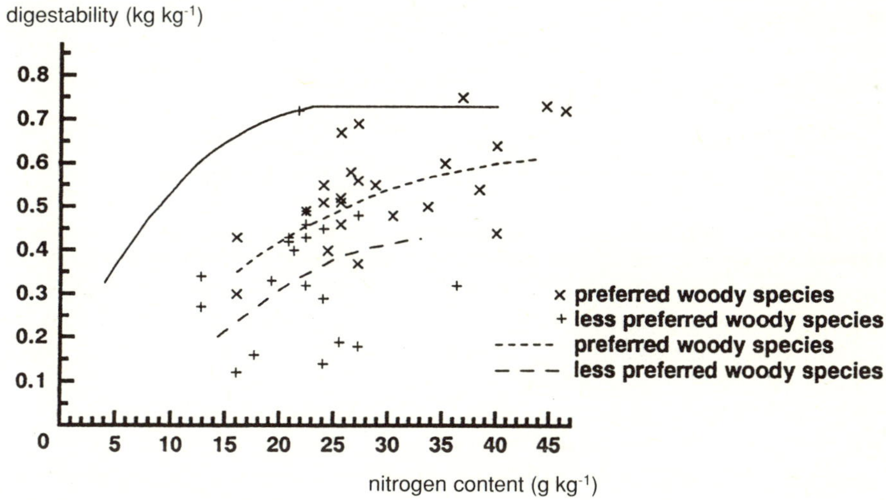

Figure 7.4: Fraction of the maximum consumable biomass in relation to the nitrogen content in the selected forage for a range of average nitrogen contents in the biomass on offer

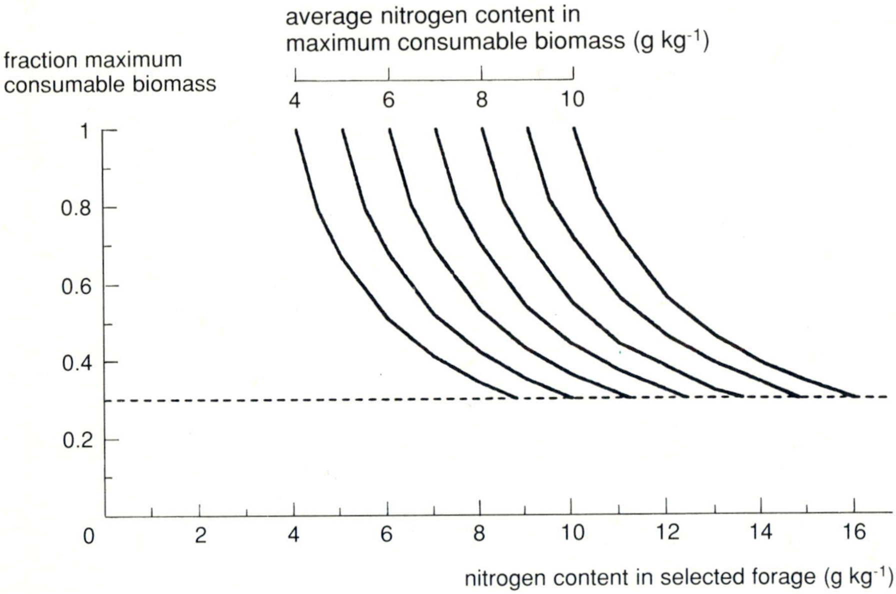

the selected forage and the amount of biomass to be consumed, determines the stocking rate. Or inversely, the production goals defined determine the intake and nitrogen content needed in the selected forage, which in its turn determines the fraction of the maximum consumable biomass which can be consumed with a corresponding stocking rate.

Available browse, its nitrogen content and the fraction which can be consumed, is treated separately. Regarding browse production, the available browse throughout the year is estimated, taking into account the maximum allowable fraction to be grazed in order to ensure sustainable growth of trees. For example, in a broad evaluation, it is assumed that 0.25 and 0.15 of all browse can be consumed in the Sahel and the Sudan respectively, but in more detailed evaluations, preference for woody species can be included as well. By treating this forage component separately, the preferences of different livestock species can also be incorporated in the evaluation.

Estimating carrying capacity

The carrying capacity of pastures, expressed in stocking rate, depends on many factors. To quantify carrying capacity, three important factors are taken into account in the evaluation method:

(i) fodder availability and its quality throughout a year

(ii) secondary production goals, i.e. the benefit from the available fodder in relation to the character or level of secondary production considered

(iii) sustainability of the production system, i.e. the fraction which can be used and still maintain future availability of fodder and its quality.

In order to compare the value of pastures, the concept of standard carrying capacity is introduced. This is defined as the maximum stocking rate based on an average rainfall year with year-round grazing, for a sedentary livestock system, at the survival level of the herd. Secondary production is limited to some meat, draught power and manure production, whereas milk production is negligible. Under these conditions, the lowest acceptable growth rate of heifers in their second year which will ensure survival of a herd, is taken as the parameter. For Sahelian cattle, cows can be expected to first conceive at a weight of 180 - 200 kg. For the survival of a herd, on average at least 3 calves have to be produced during the life-span of a cow (half males, half females, and 0.3 mortality). Assuming an inter-pregnancy period of two years and a maximum life-span of twelve years, the first calf has to be delivered in the sixth year. Calves reach a weight of 80-100 kg by the end of their first year. During the following four years heifers should grow 100 kg to reach a weight of 180-200 kg in their fifth year (first conception), implying the growth rate of a heifer as being at least 25-30 kg per year. This production goal can be translated in terms of the minimum required nitrogen content and related intake rate needed per heifer, which can subsequently be expressed as the intake rate per TLU. Using the intake model, it is calculated that such an intake rate corresponds to 5.6 kg DM d^{-1} TLU^{-1}, with a nitrogen content of at least 7.5 g kg^{-1} during the dry season.

An example of estimated carrying capacity

As an example, the standard carrying capacity (expressed as animal biomass in ton per km^2) per rainfall zone is presented in Fig. 7.5. In this figure, 0.35 of the estimated biomass production of the herb layer per zone (Fig. 7.2) is regarded as the maximum consumable biomass under year-round grazing in a sedentary system. Quality is assessed by combining the average nitrogen content of the biomass on offer in the dry season and the nitrogen content in the selected forage, which determines the fraction of the maximum consumable biomass which should be consumed (Fig. 7.4). Browse production per zone and its nitrogen content is estimated to determine the available biomass, in addition to the consumable herb layer. Consequently, the calculated animal biomass per km^2 should comprise grazers and browsers. The line drawn in Fig. 7.5 corresponds with a quality criterion expressed in a minimum nitrogen content of 7.5 g kg^{-1}.

In the same figure, actual stocking rates in Niger, Mali and Burkina Faso in 1982 are presented as well. It appears that eight years after the severe drought in the early 1970s, standard carrying capacity in the northern and southern Sahel had been re-attained and in some areas even exceeded. Since 1982, it is

Figure 7.5: Animal biomass as function of the average annual rainfall per year

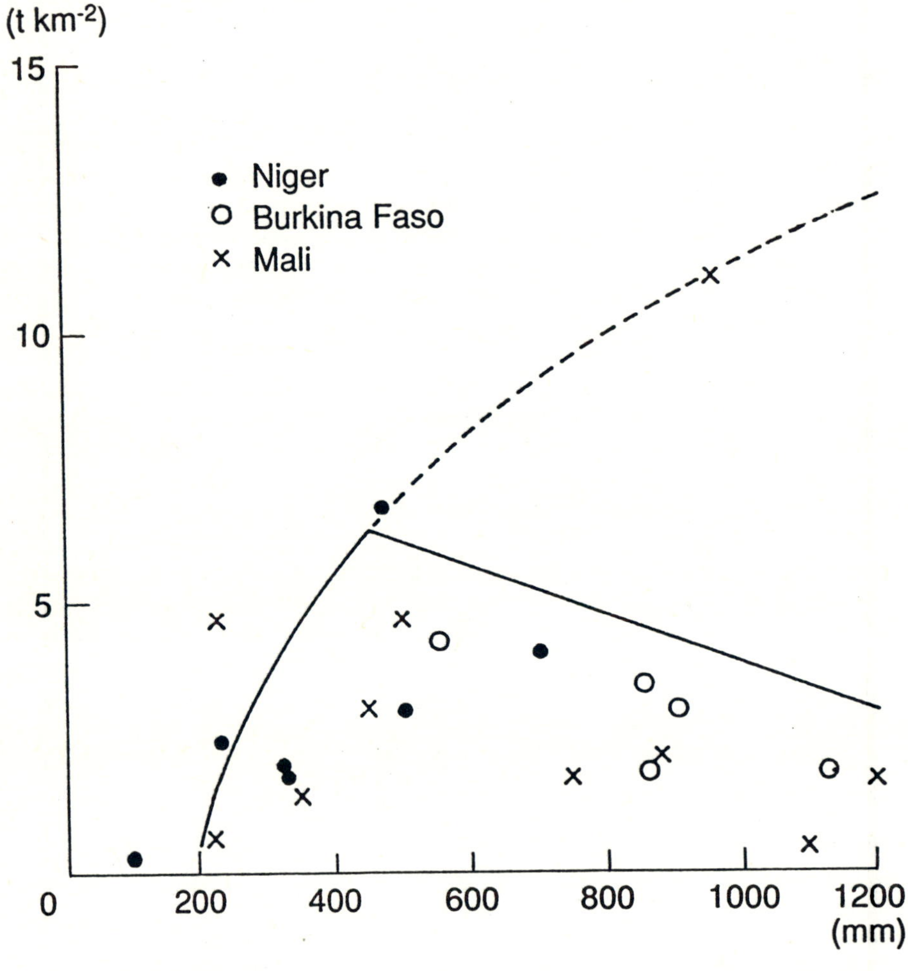

Note: The solid line represents the standard carrying capacity (see text); the interrupted line does not include quality of the forage. Individual observations concern estimated stocking rates in 1982 in different parts of three Sahelian countries.

confirmed (e.g. Bourn and Wint, 1986) that stocking rates in Niger exceeded the standard carrying capacity. The indicated data for the northern and southern Sahel in Niger represent the average values per department. However, a detailed aerial survey of a region (81.550 km^2) north of Tahoua produced a figure of 1100-1500 kg km^{-2} (standard carrying capacity being 550 kg km^{-2}) during the period of 1981/1982. In September 1985, after the extreme drought of 1983/1984, this figure dropped to 330 kg km^{-2}. In the Sudan zone, actual stocking rates are generally lower than the estimated standard carrying capacity, except for the cotton-growing region in the south of Mali (indicated by the cross on the interrupted line in Fig. 7.5). This is possibly explained by the availability of cotton seed as concentrate, which allows a higher fraction of the available fodder to be consumed.

Estimating carrying capacity in relation to production goals

Although standard carrying capacity is a satisfactory tool for comparing the value of pastures, it is of more practical interest to relate carrying capacity to production goals and production systems. The evaluation method provides for this option. As an example, carrying capacities for three regions in the south of Mali, covering three different production systems and two production goals, are estimated and compared to actual stocking rates (Table 7.1).

The three regions, Kadiolo, Sikasso and Koutiala, differ strongly in character: average annual rainfall declines from 1400 mm in Kadiolo via Sikasso to 800 mm in Koutiala; cultivated land comprises 0.03, 0.11 and 0.23 of the total surface of Kadiolo, Sikasso and Koutiala, respectively; and in the same sequence, population pressure amounts to 21, 28 and 29 people per km^2.

Table 7.1: Estimated carrying capacity (ha TLU^{-1}) of three regions in southern Mali; assuming three different systems and two production levels, compared to (actual) stocking rates (ha TLU^{-1}) observed in 1986 (for explanation see text)

Livestock system (menu)	*Production system*	*Carrying capacity*		
		Kadiolo	*Sikasso*	*Koutiala*
1	25	7	11	14
	50	8	14	17
2	25	6	7	7
	50	8	9	8
3	25	6	6	5
	50	7	8	7
stocking rates		10	6	4

Source: Leloup and Traoré (1989)

Cultivation of cotton is negligible in Kadiolo, but very important in Koutiala. The production systems to be considered can be described as three different types: Type 1 represents a system in which the natural pastures are used by livestock, which has no access to the arable land residues nor to agricultural by-products. In Type 2, livestock has access to arable lands, but does not receive agricultural by-products of higher quality, as would be practised by arable farmers in sedentary livestock systems. In Type 3, all available good quality by-products are used, such as cotton seed and the hay of beans and peanuts. Production levels are expressed by the weight gain of heifers in their second year. The first level, 25 kg yr^{-1}, represents the situation indicated earlier in which herds do not grow and mainly produce draught power and manure. The second level, corresponding to a weight gain of 50 kg yr^{-1}, characterises the Sahelian livestock system, with a level of production of meat and milk sufficient for a pastoral life style.

From the results in Table 7.1, it appears that the standard carrying capacity (Type 1; 25 kg growth yr^{-1}) decreases from the south (Kadiolo) to the north (Koutiala). This is due to the decreasing importance of arable farming from south to north; arable farming competes with the best grazing grounds, leaving only the marginal grounds available for grazing. If the aftermaths of arable lands can be used for grazing (Type 2; 25 kg growth yr^{-1}) and agricultural by-products are used (Type 3; 25 kg growth yr^{-1}), carrying capacity increases in Sikasso and Koutiala, whereas the situation does not appreciably change in Kadiolo, the region with the lowest rate of cultivated land. To obtain a higher production level (50 kg growth yr-1) stocking rates have to be reduced in all cases, but to lesser extent when agricultural by-products of good quality are available (see Sikasso and Koutiala, Types 2 and 3).

When actual stocking rates are compared to estimated ones, it appears that production goals in the three regions are presumably different. In Kadiolo, production of meat and milk is possible at the actual stocking rate and could be even higher than the assumed 50 kg growth yr^{-1}, since there is still an opportunity for selection of the best forage. However, in Sikasso there is only enough fodder available (Type 3) to maintain herds for the production of manure and draught power in support of arable farming. The stocking rate in the Koutiala region is in any case exceeded.

Conclusion

The need for a thorough knowledge of Sahelian rangelands stems from the negative changes that have occurred in animal husbandry in this region, where animal husbandry is particularly dependent on fodder, determined by the natural rangelands.

The majority of the rangeland area is dominated by annual species, due to the extreme climate in which a short rainy season is combined with a high evaporation/rainfall ratio. This climate is reflected in strong vegetation dynamics, and a high degree of sensitivity by perennials to different forms of stress.

The generally poor soil implies that only in the northern Sahel is production limited by water availability, while elsewhere in the region nutrient availability is the limiting factor. Consequently, rangeland quality is high in the northern Sahel and low in the rest of the region, leading to a high degree of mobility in livestock systems.

Special methods have been developed to tackle the particular problems encountered in evaluating these rangelands:

(i) the evaluation is based on soil and climatic properties rather than on vegetation characteristics

(ii) the approach for estimating production limited by water is different from that used for estimating production limited by nutrients. The same applies to annual and perennial species

(iii) field observations are used to verify the theoretical evaluation; field observations in a given year are translated to normal and dry years

(iv) animal production and carrying capacity is estimated for both mobile and sedentary livestock systems.

The usefulness of these methods is illustrated in three case studies.

8
Biological Sustainability in Pastoral Systems: The Kwazulu Case

David Tapson

> Asking an extension officer to design the optimum grazing system is like asking an amateur engineer to design an economical bridge of undefined span to carry an unspecified but variable load; built from materials of unknown specifications; joined by unreliable techniques in an environment conducive to advanced structural deterioration of an ill-defined nature (Higgins, 1967).

Introduction

The Zulu people occupy an assembly of ten geographically variable blocks of land, which together comprise KwaZulu, designated a self-governing homeland under the 'separate development' policy of South Africa, now rapidly becoming an historical aberration. At present, Kwazulu comprises some 3.2 million hectares, situated on the east of the subcontinent, intermingled with Natal between the Drakensberg escarpment and the Indian Ocean coast. It has Mozambique as a northern neighbour and extends to approximately 31^0 south of the equator. Livestock, in particular cattle, are a major feature of the Zulu rural economy, and it has been argued for decades that the livestock/grazing system is threatened with collapse, as a consequence of over-utilisation of the grazing resource. Within the general concept of sustainability, which is dependent on issues of equity, viability and ecological soundness, this chapter deals with the last-named, i.e. the biological sustainability of the pastoral system existing in KwaZulu.

Any animal production system comprises essentially three subsystems: the food resource exploited by the animals; the animals and their internal reproduction dynamics; and the interaction between the two which is controlled by the management regimes applied to each of the other subsystems. In the case considered here, concerning as it does an essentially unsophisticated production system, the food resource employed is unimproved veld grazing. The animal sub-system is the total cattle herd owned jointly by a large number of smallholder farmers, and the management sub-system is that set of practices which those owners apply to both the grazing and cattle resources. The system includes the added complication arising from individual ownership and therefore control of the cattle, linked to common ownership of the grazing resource, with all the limitations to individual choice that this implies.

The grazing component has particular importance arising from the fact that degradation of the resource and associated soil erosion are cited as sufficient reason for intervention to change the nature of cattle ownership in KwaZulu, indeed in almost any of the less developed areas of the sub-continent. With the wealth of comment on the degree of overstocking in force in KwaZulu, policy and strategy proposals, as may be expected, tend to take the necessity for destocking as given. Analysis of data derived from economic data shows clearly the destructive effect this action would have on the welfare of cattle owners. The effects are such that destocking should not be embarked upon without thorough analysis to determine whether the welfare cost can be justified by the predicted environmental benefit. This chapter considers first the value of the basic grazing resource, and then questions the degree to which it has in fact been degraded as a result of grazing malpractices. This brings into question the very foundation of the argument for destocking. The arguments presented are crucial to both the sustainability of the cattle sector in KwaZulu and the issue of equity in the system.

The grazing resource

KwaZulu is an area of high agricultural potential, by the standards of the southern African subregion. The *Preliminary Land-use Plan for KwaZulu,* published by the KwaZulu Department of Agriculture and Forestry (1980:29-49), describes, quantitatively and in detail, the potential of the ten homogenous farming areas of which KwaZulu is comprised. In short, of KwaZulu's total land area of 3.32 million hectares, some 2.24 million or 68%, is described as 'natural grazing'. Of this area, 28% falls into the high potential zone, 20% into the medium potential zone and 51% into the low potential zone. A further 411,600 ha of high potential arable land, would also be available as grazing in the winter. In total therefore, 1.50 million ha out of 3.32 million would provide grazing of high or medium potential. While the 1.82 million ha is described as 'low potential' on account of its inherently low carrying capacity, it does have the advantage of providing high quality grazing and browse if managed correctly (KDA, 1980:45,47,48). This means that high levels of animal nutrition can be maintained throughout the year from the veld, without recourse to purchased or cultivated forage supplements.

Tainton (1981:Fig. 2.1) shows that KwaZulu is dominated by Fire Climax Grasslands of potential forest and savanna areas, regularly interspersed with intrusions of Valley Bushveld. In the north, patches of Fire Climax Grasslands of potential savanna areas, and Eastern Mixed Bushveld occur. For management purposes, these classes of grassland are reduced to the groupings: sweet, sour, and mixed veld. KwaZulu falls almost entirely into the sourveld category, with minor portions of sweet and mixed veld scattered throughout the territory (Tainton, 1981:Fig.2.3). Sourveld is characterised as having rapid and regular growth as a consequence of relatively high rainfall, and therefore a higher carrying capacity than the sweetveld. The sward is described as dense and capable of withstanding moderate levels of overgrazing. Such treatment does lead to reduced productivity, due more to changes in species composition than reduction in cover (Tainton, 1981:45).

In sum then, the grasslands of KwaZulu constitute a substantial grazing resource, capable of delivering some 23% of the region's net farm income (KDA, 1980:69). It is, however, an article of faith that this valuable resource has been seriously degraded as a consequence of past overgrazing and mismanagement, and that the process continues today. As a result, it is argued that production from the grassland is considerably lower than is potentially possible, and based on the continuing degradation, is likely to decrease in future. Since the absolute limit of production from the animal system is determined by the primary productivity of the underlying grazing, an understanding of the grazing system dynamics is necessary to assess the implications of this for the whole cattle production system. The following section will therefore provide a brief outline of the theoretical basis for determining the productivity of grasslands, as a precursor to testing the assertion that the productivity has in fact declined.

Grassland dynamics and the KwaZulu case

Models of grassland productivity are predicated on the Pearl-Verhulst or logistic equation of theoretical biology. This is a quadratic of the form

$$dx/dt = Ax(x-\bar{x})$$

where A is a constant, t is time, $\bar{x}$ is the equilibrium, unexploited population and x the population at any given time. A model widely used in South Africa (Tainton, 1988:4) based on this relationship, is that developed by Jones and Sandland (1974:335). Using experimental data, they showed that gain per animal is linearly related to stocking rate, thus:

$$Y_a = a - bx$$

where a and b are constants, Y_a is the gain per animal and x is the stocking rate (animals per hectare). Where this holds, the relation between stocking rate (x) and gain per hectare (Y_h) is a quadratic of the form

$$Y_h = ax - bx^2$$

where a and b are constants.

They show further, by differentiation, that maximum gain per hectare occurs when the stocking rate is half of that where $Y=0$, at nmax in Fig. 8.1. Using data from a wide range of experiments they show the generality of these relationships, illustrated with data from a tropical grass-legume pasture, grazed continuously by beef cattle (Fig. 8.1).

The foregoing provides the three relationships essential to critical understanding of the dynamics of overgrazing and resource degradation. These are:

(i) Gain per animal will decline linearly as stocking rate increases, and can be negative if the limits set by the population dynamics of the grazing resource are exceeded.

Figure 8.1: The relationship between gain per animal and gain per ha in response to increasing stocking rate

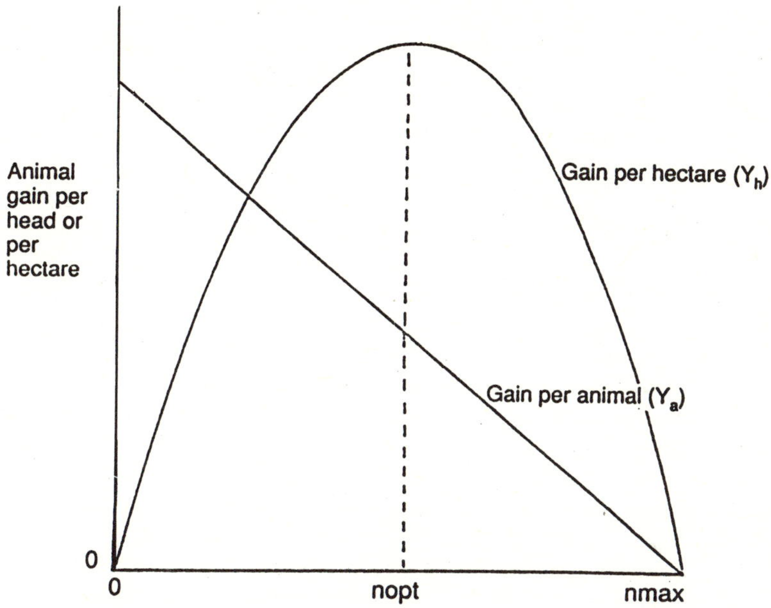

Source: After Jones & Sandland (1974:337)

(ii) Gain per hectare proceeds quadratically as stocking rate increases, with two equilibrium points at zero. The first of these is when stocking rate is zero, and the second occurs at that stocking rate at which gain per animal has declined to zero, at 'n max' in Fig. 8.1.

(iii) Maximum gain per hectare occurs at the midpoint between these two values, 'n opt' in Fig. 8.1. This is the point of 'maximum sustainable yield' (MSY), and it follows that mathematically, stocking rate can never exceed twice the theoretical MSY level. This is the basis upon which all comparisons of stocking rates are made.

Sustainability of the KwaZulu grazing resource

The implications of the relationships above need to be brought to bear on the long history of statements about impending catastrophe in the grasslands of KwaZulu. The earliest traceable reference to the problem in KwaZulu is in a document entitled 'Planning 27-12-38', quoted by Franklin (1948:128): 'The persistence of native cattle holding... is even more hampering to economic development, because religious cattle cumber and overstock the earth...' Brookes and Hurwitz (1957) refer to stock maintained at starvation point through the evils of overstocking and note that measures such as compulsory destocking and veld management schemes have had little effect in improving the veld. The litany continues with Thorrington-Smith et al.(1978:93) referring to the veld being brought to a state of near disaster as a consequence of overstocking; KDA (1980:85) estimating over-stocking to be of the order of 100% (which is at least mathematically possible) and the policy statements of the Minister of Agriculture (1977-80) exhorting farmers to decrease livestock population and to conserve the natural resources of KwaZulu.

The image projected, and one firmly held by planners and policy makers, is that the grazing resource of KwaZulu is heavily overstocked, by some 100-300%, and that severe degradation of this resource has occurred as a result. As a logical consequence, the supposed degradation of the grassland is deemed to have caused soil erosion on an unacceptable scale. These two conditions are held to be threatening the sustainability of cattle production on environmental grounds and are advanced as sufficient cause for stringent intervention by the authorities to halt the process, i.e. based on the 'ecological' objectives referred to by Sandford (1976:3). With sustainability at the root of the argument for intervention, the validity of the received wisdom regarding it requires examination.

The arguments presented in this section will fall into two parts, to support the contention that, assembled as a body of evidence, the information available is too weak and incomplete to justify any intervention, including an action as drastic as destocking to the levels held by technical specialists to be 'optimum' stocking rates. The first part of the argument presents historical evidence which suggests that alarmist statements about the state of the grazing have been premature or possibly ill-informed. The second part examines some contemporary research findings which show that, at the least, one should be cautious of the assumptions upon which these alarmist statements are based.

Historical evidence

The essence of the warnings made over the last half century is that the grazing resource is at an advanced stage of destruction and that as a consequence, soil erosion is proceeding at an unacceptably high rate. The primary cause of this is held to be the degree of overstocking the veld has suffered, and continues to suffer. As was shown above, the theory underlying the estimation of carrying capacity is rigorous, and suggests a degree of precision is required in making estimations. Estimates that the stocking rate is up to three times greater than the optimum based on carrying capacity are therefore, on the face of it, nonsensical. They can be explained only by errors in estimating either the carrying capacity, or stocking rate, or both. It could be argued that this is an over-precise and pedantic viewpoint, and that regardless of the mathematical niceties, the effect of the pressure on grazing caused by the stocking rate (whatever it is), is causing demonstrable degradation which will inevitably lead to an irreversible breakdown.

The references quoted above show that this view has been extant for some fifty years, and the very length of this history provides the opportunity to

Figure 8.2: Herd count and mortality in the KwaZulu herd, 1974 to 1988

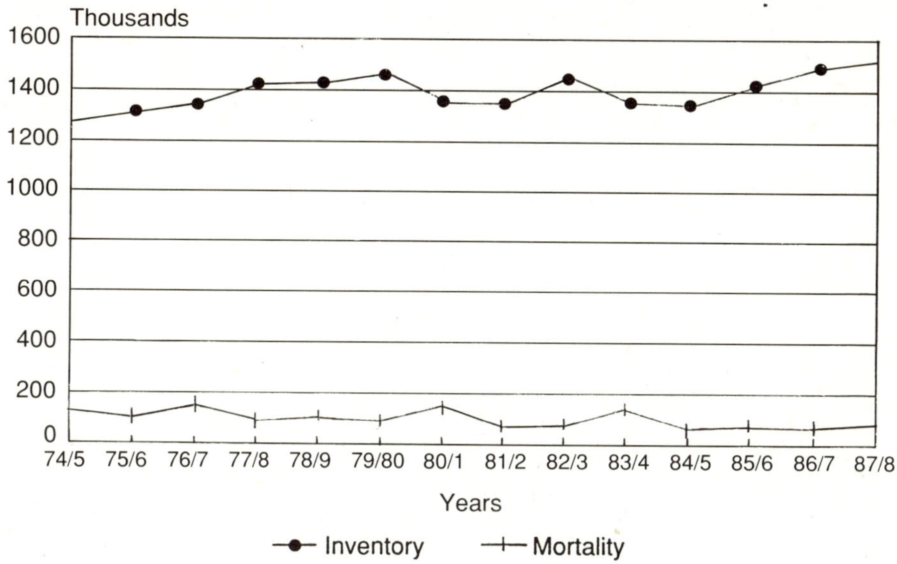

Source: Annual Reports of the Director of Veterinary Services

derive omnibus or holistic indicators of the validity of the statements. Had they been true over all or even part of this time, the decline in primary productivity, as a consequence of the collapse over time of the basic resource, would have resulted in a decline in stock numbers, accompanied by a decrease in offtake and an increase in mortality. The evidence suggests quite the reverse. The stock census maintained by the self-governing administration of KwaZulu since 1974 is probably no more accurate in absolute terms than any other developing country, but it has been recorded systematically and consistently. The time series also coincides with the period over which KwaZulu's land area has been relatively stable. The records show that stock inventory has increased from 1.27 million in 1974/75 to 1.515 million in 1987/88, a peak resulting, no doubt, from a series of satisfactory growing seasons. Over this time period the inventory has fluctuated about a mean of 1.39 million with a coefficient of variation of 0.05. Over the same time, sales, with a mean of 0.8% and mortality at 6.7%, have fluctuated far more widely, neither showing a trend over time which would indicate a deterioration in primary productivity. Mortality has in fact shown a declining trend (Fig. 8.2).

It must be reiterated that this fifteen year time series comes at the end of a fifty year spell, over which period it has been claimed that the veld has been facing an ecological catastrophe. In the face of similar data in Zimbabwe, Sandford (1982a:57, quoted by Cousins, 1987:72) enquired whether such prolonged defiance of the 'Laws of Nature' could not be accounted for by under-estimation of the initial carrying capacity.

The validity of the assertion that deterioration has occurred due to overgrazing, or has occurred at all, is specifically challenged by Fowler (1981) in Swaziland and Sandford (1982a and 1983) in Zimbabwe. Using regression functions to analyse death rate, extraction rate and calving percentage, and comparing these with rainfall, Fowler (1981:7-10) concluded that the production statistics showed no significant trends in the period 1954-64. However, between 1967 and 1977 all three rates showed significantly decreasing trends. This could not be attributed to rainfall variations between the two periods as there was no significant difference. He concluded that while the data presented supported current claims of overstocking, it was of more recent occurrence than is generally supposed and claims of overstocking made in colonial times were probably not valid.

Sandford (1982a:43-45) argues strongly that the normal definition of 'degradation' suffers both from inexactitude and difficulties in measurement. He develops this argument further (Sandford, 1982a:13-19) putting the case against what he terms the 'Mainstream View'. He maintains that the causes of degradation and desertification are so many and variable that to attribute them solely to the effect of 'overstocking' is not supportable. In this he is supported by Johnson (1979:8), who cautions that influences like the invasion of agriculture into grazing land, and the concentration of stock about water points must also be considered. Further, the causes of gulley erosion are 'discipline-specific' and can be variously attributed, depending on whether the observer is an agricultural scientist, geologist, palaeobotanist or archaeologist (Cooke and Reeves, 1976:6, in Johnson, 1980:9). It must be noted that in arriving at his conclusions, Sandford does not use the very substantial body of research on grasslands which has been built up in southern Africa since 1960, the bulk of which is reported in the Journal of the Grassland Society of Southern Africa.

Both Southey (1981:67) and Sandford (1982a: 7, quoted by Cousins 1987:72) hold that in any event 'optimal' carrying capacity is a value judgement. It would vary according to the objective of the range manager, for example even for so simple a differentiation as between output per animal or output per hectare. This concept suggests an interesting comparison for Kwazulu. The survey data indicate that useful offtake from the sample herd, that is including sales and slaughter for home consumption, was 7.1%. Extrapolating this offtake to the long-term mean inventory of 1.39 million implies an offtake of some 99,000 head. Accepting that the carrying capacity of KwaZulu is of the order of 465, 000 LSU's (Large Stock Unit) and that the average LSU value for a herd of the type in KwaZulu is 0.75, the 'optimal' herd size would be some 620,000 head. The offtake from a ranch herd run on an extensive basis with levels of management which could reasonably be assumed for KwaZulu has been estimated at 15.6% (Tapson, 1982:5). Applying this offtake coefficient to the optimal herd size would give a total offtake of approximately 96,720 head. If one then includes the increased benefit from milk, bridewealth (*lobolo*) and draft in the larger herd of KwaZulu, even allowing for the crudity of the estimating procedure, it seems at least possible that present stocking rates are capable of delivering a greater stream of benefits than the 'optimal' rate.

In summary, the historical evidence suggests that anxiety over the impending collapse of the ecosystem was exaggerated or at least premature (see Fowler, 1981). To date at least, the ecosystem in KwaZulu has stubbornly resisted complying with the prediction of imminent collapse. Whether it will continue to do so or not, depends upon the inherent stability and resilience of the grassland system. This is a technical attribute, which can be assessed with some degree of reliability even though, as may be expected, conflicting views of both the meaning of the terms, and their application to KwaZulu, can be found. Some of these viewpoints are considered in the next section.

Technical evidence

Mentis (1985:27) fires a philosophical broadside at the practice and principles of 'conservation', as applied to rangeland resources. He refers to a definition of conservation he attributes to the World Conservation Strategy (WCS, 1980, quoted by Mentis, 1985:27), namely 'the management of human use of the biosphere so that it may yield the greatest sustainable benefit to present generations while maintaining its potential to meet the needs and aspirations of future generations'. Few would quarrel with that. Mentis (1985:27) describes it as being altruistic, expressing only one of many guidelines, being value-based, and finally being non-falsifiable in the sense that it is not amenable to tests which can 'show incompatability of observation with explanation, expectation or prediction'(Mentis, 1985:27).

Mentis (ibid.:28) argues with Hardin's position (1968:1246) that altruism is genetically self-eliminating through Darwinian selection, and that since observations and measurements rest on value judgement, facts are theory-laden, and value-neutral knowledge is unobtainable. He accuses 'rangeland thinking' of venerating the climax and the stable condition this implies. On the contrary, since a condition of disequilibrium can be more productive than the climax condition, the challenge of management is in fact to hold the system in an appropriate state of disequilibrium (Mentis, 1985:29). He holds that the conservation concept is undermined through its dependence on empty words such as 'optimal' which can only be defined within the framework of a specific value system. He concludes that, as currently defined, conservation is hardly operational and its success defies evaluation (Mentis, 1985:30).

This simplified critique of conservation serves as an introduction to the concepts of stability, resilience, and optimality which are central to the issue of sustainability of the KwaZulu grazing resource. It also serves to question the applicability of grassland research findings to the practice of grassland management.

Walker (1980:79) provides an analytical model which contends that in semi-arid savannas (undefined) the result of good veld-management aimed at high sustained production is to reduce the resilience of the veld. He defines a stable system as one which does not change readily in response to stress but, if the stress is sufficient to cause a critical degree of change, may well not return to a condition of equilibrium. A resilient system, on the other hand, may well not be stable, but may be changed considerably and still return easily to its equilibrium state. His model showed that heavy grazing favoured all the features leading to a high level of resilience. Alternatively, under 'good management', species changes occur leading to a lower level of resilience. He remarks on the fact that the tribal grazing areas of southern Africa have been reported to be on the brink of collapse for 50 years, yet continue to persist, despite localised disasters due to periodic drought. On the other hand, neighbouring well-managed ranches with higher production and more stability suffer far more devastating effects when subject to the same drought (Walker, 1980:83).

Walker's (1980:79) theoretical analysis suggests that heavily grazed, 'degraded' systems, typified by the KwaZulu case, are inherently more resilient than climax systems, and therefore arguably more sustainable. In the case of Transkei, on the border of and very similar to Kwazulu, McKenzie (1982:21) supports this finding empirically. He found that, despite stocking rates held to be 2.5 times higher than the acceptable carrying capacity (cf. above), much of the grassland contains an abundance of climax species and has a high basal cover. 'Such features tend to contradict the expected consequences of such heavy utilisation...'(McKenzie, 1982:21).

This quotation is supplied as an introduction to a theme that persists through the bulk of reports on research in grasslands - the stubborn refusal of the ecosystem to fulfil the predictions of those that research it. Other examples will be reported, and by themselves provide a notable reason for caution in proceeding with interventions. McKenzie (1982:22) elaborates the stability concept to include resistant stability and recuperative stability which is close to Walker's (1980:79) resilience concept. McKenzie (1982:22) defines resilience as the ability of a system to absorb changes in state variables, and still persist intact through time. The particular effect of these concepts in Transkei, is that the desirable climax *Themeda* veld, being lower on the resilience scale than the less desirable *Aristida*, has been largely replaced by it. The *Aristida* on the other hand is showing both high resilience and resistant and recuperative stability, and appears unlikely to be replaced by any other system, either higher or lower in the succession. In other words, it is a degraded but highly sustainable condition - supporting in addition a stocking rate more than twice as heavy as its 'accepted' carrying capacity (McKenzie, 1982:21).

Trollope (1990, pers. comm.)[1] has pointed out that Walker's (1980) and McKenzie's (1982) arguments remain untested hypotheses and that personal observation of very well-managed climax veld following the 1982/83 drought indicated that it recovered considerably more rapidly than heavily grazed veld. Danckwerts and Stuart-Hill (1988:218) however, had different results from observations of veld during and after the drought. They used the classification of species used by grassland scientists to assess veld condition. For this purpose species are classified as Increaser I which dominate in poor veld and which increase when it is understocked or selectively grazed; Increaser II species which dominate in poor veld and increase with overgrazing; and

1. W.S.W. Trollope, Professor and Head, Sub-department of Plant Science, Private Bag X1314, Alice 5700, South Africa

Decreaser species which dominate in good veld but decrease when it is mismanaged (Trollope et al., 1990:52).

They found that Increaser I species suffered the least mortality during the drought, followed by Decreaser then Increaser II species with the highest mortality. In contrast however, the ability of these grasses to recover after the drought followed the reverse order (Danckwerts and Stuart-Hill, 1988:222). This seems to be consistent with the hypothesis of Walker (1980:79) and McKenzie (1982:21), that heavily stocked ('overgrazed') veld is inherently more resilient, able to recover and therefore persist, than moderately stocked (and better managed) veld.

Applicability of the stocking rate model

The evidence presented to this point has been intended to show firstly that a concept as basic as conservation is value-loaded. This being so, concepts which flow out of it, such as optimal utilisation rate, are equally value-loaded, and depend very largely on the objective function of whomever controls the resource. Secondly, whatever the value system which governs decisions on utilisation, it appears that the degraded, heavily-stocked systems are inherently more resilient than veld closer to the climax condition. Yet to be considered is the applicability of the stocking rate model as a predictor of grassland behaviour or as a management tool. Evidence will be presented to suggest that, despite its apparent mathematical precision, the model's performance even under tightly controlled experimental conditions is such that as a policy guide, the model should be approached with caution in a situation where human populations will be profoundly affected by the application of that policy.

Over the past two decades a considerable body of knowledge has built up as a consequence of research into the maintenance and productivity of grasslands. It is proposed here to select pieces of this work to illustrate that the basic tools of the grasslands research process are less than adequate for making critical decisions involving the welfare of very large groups of people. Research into grasslands is aimed at determining optimum veld management practices, the efficacy of which is measured by the output of animal mass from grassland being tested. Since the concept of 'optimum' includes maintaining the condition of the veld on which the animals graze, measures of veld condition have had to be derived as well. The basic tools of research in grasslands science are firstly, the stocking rate models of Jones and Sandland (1974) and derivatives of this original model. Secondly, various techniques of veld condition scoring are used to monitor changes in the quality of the grazing resource, and to determine the mean grazing capacity of the grazing site, based upon the supposition that the different species available have specific forage production potential (Tainton, 1981:78).

Typically a research programme will include a number of treatments based upon variations in the stocking pressure and time of grazing. The effects are analysed and evaluated according to the Jones and Sandland (1974) stocking rate model, while the veld condition changes occurring for each treatment are measured using the veld condition scoring techniques.

The Jones/Sandland model, based as it is upon the universal logistic growth curve, has a defensible theoretical base, to the extent that it can be used to measure the performance of any biological population including the sward of a grassland. In practice, however, the performance of that grassland is largely expressed via the biomass of the animals utilizing it, which introduces a complicating remoteness from the theory. On the other hand, the veld condition scoring techniques which are used not only to monitor the condition of the veld but also to predict their grazing capacity (Tainton, 1988:78), suffer from imprecision and a modicum of value judgement. The basic procedure of the veld condition score is to determine the proportions of key species in the site under examination and to compare these with the proportions occurring in what is described as a 'benchmark site'. The grazing capacity is determined by estimating the forage-producing potential of the combination of species in the test site, and then prorating them with the forage-producing capacity of the benchmark site. Assuming that the grazing capacity of the benchmark site is known, a mean grazing capacity for the test site can be determined by simple proportion (Tainton, 1988:78). The selection of the benchmark sites is described by Tainton (1981:47) as follows:

> The identification of benchmark sites is subjective. It involves the selection of sites which are productive and stable, and represent the type of veld which one suspects was widespread prior to large scale settlement and veld degeneration in South Africa.

The veld condition score techniques are subject to a number of criticisms. Firstly, scoring is a subjective procedure based upon the selection of a benchmark site, which is essentially a researcher's perception of a condition which might have existed, but which he patently has never seen. Secondly, while in the more stable humid grasslands benchmark sites might retain their characteristics over time, it is not so in the areas of lower and more erratic rainfall. Because condition of benchmark sites will vary both within and between seasons, the site requires a re-assessment each time the test sites are analysed (Tainton, 1981:55). Thirdly, the value of the climax grassland or stable state, which is the basis of the reference benchmark site, is completely subjective and highly dependent upon the objective function of the manager of the grassland. All these criticisms notwithstanding, the procedures adopted in the process of measuring veld condition are mathematically rigorous and have been fully tested by a number of researchers over a wide range of conditions. The procedure therefore cannot be faulted, although it is conducted within a number of questionable assumptions which are, in the end, the basis of the judgements made from veld condition assessment scores.

Hardy and Hurt (1989:51) evaluated six veld condition assessment techniques in the Highland Sourveld of Natal. The data were drawn from an existing stocking rate trial which had been in operation from 1977 to 1988. Several stocking rate treatments were applied, using accepted veld management practices, including multi-camp systems and rotational grazing. The six veld condition scoring techniques which were evaluated cover the full range of those commonly in use and included specifically the techniques recommended by Tainton (1988:78) for the grassveld areas. The treatments included stocking pressures varying from slightly lighter than the potential grazing capacity of the Highland Sourveld to pressures of the order of two and a half times this figure. The range of treatments therefore provided a basis from which the effect of stocking rate on proportional species could be evaluated over a relatively short period of time.

From the viewpoint of this discussion, the results are interesting more for what they do not show than for what they do show. The researchers conclude that none of the techniques tested were sufficiently sensitive to index veld condition, although the two methods recommended for the grassveld areas (known as the key species methods) were the most sensitive. However, according to these methods, *all* the stocking rate treatments caused a decrease in veld condition over time, despite the very wide range of pressures between treatments. The researchers go on to note that what was not considered by the techniques is the correlation between loss of the key species and accelerated soil loss, or even whether they indicated a significant reduction in long term animal production. The conclusions are worth quoting:

> Methods of indexing veld condition presently applied in the humid grasslands appear to be insensitive to grazing intensity and should not be used for monitoring veld condition. The use of these methods for estimating grazing capacity (where the method is based on proportional species composition alone) is unlikely to lead to an improvement in veld condition and could considerably lead to further deterioration of veld. The key species method appears to have the greatest potential for indexing veld condition and future research should be aimed at identifying those species which react most sensitively to grazing management (Hardy and Hurt, 1989:57).

The implications of these findings need to be borne in mind when considering the results of two research reports by Turner and Tainton (1989a and 1989b). The first of these (Turner and Tainton, 1989a:175) measured the relationships between herbage mass, veld condition, stocking rate and commencement of spring grazing on the one hand, and animal performance in relation to herbage mass, stocking rate and initial animal mass on the other hand. The trials were carried out in the Tall Grassveld area of Natal. The procedures were routine and of the type commonly accepted in grasslands research. Veld condition was estimated using both productivity methods and the key species methods, with the latter being modified for reasons not explained. The models developed in terms of herbage mass, veld condition and stocking rate were internally consistent and showed, not surprisingly, that at a particular stocking rate the residual herbage mass at the end of the season increased with veld condition, and on the other hand decreased with stocking rate, at significant levels in both cases.

Patterns in animal production, however, were somewhat unexpected. In particular, there were no significant linear relationships between average daily gain and stocking rate at any of the sites, and therefore no Jones and Sandland model (1974) could be constructed for animal performance. In all cases however, at the end of the season those animals grazed at the more moderate stocking rates had shown the most gain. The authors suggest that animal performance only declines linearly once the available herbage decreases to such an extent that intake of herbage becomes the limiting factor to animal performance. This suggests further that the heavier stocking rate used in the trials did not reduce the herbage available to this point. A more surprising result was the fact that generally poor animal performance was obtained on the sites with the best veld condition scores. The model developed was highly significant and showed that average daily gain was negatively related both to stocking rate and herbage mass, but positively related to the initial mass of the animals going into the trial. It should be noted that, since herbage mass is positively related to the veld condition index, the implication is that animal performance is negatively related both to herbage mass and veld condition. However, these results cover only a single summer and longer term trials might vary the results substantially.

These results need to be contrasted with those presented by the same authors (Turner and Tainton, 1989b:183) who tested animal performance in relation to herbage mass and stocking rate in the Natal Sour Sandveld. The methodologies and techniques were basically identical to those used in the trial reported on above. In this case, however, there was a significant linear relationship between average daily gain and stocking rate over a 195 day period, which was predictable on the basis of the Jones and Sandland (1974) model. The grazing system, however, did not completely give up its independence, as animal performance declined more rapidly with stocking rate at the site which had in fact enjoyed the higher rainfall.

The effect of rainfall variability

These results are provided to do no more than illustrate the as-yet unexplained complexity of the interrelationships between rainfall, the grazing system and the animal system which exploits it.

The principal effect of rainfall is the complications it introduces by virtue of the variation it causes in yield of the grazing system from year to year. The managers of the cattle using the grazing system would prefer to have a relatively stable and continuous flow of fodder. The degree of variability experienced is illustrated by Booysen's work (1981:67) in the Tall Grassveld of Natal, near Pietermaritzburg. By comparison with the veld types associated with the lower and more erratic rainfall patterns, this veld type is considered to be relatively stable. However, Table 8.1. shows that the inter-seasonal difference in control plot yield not only exceeded the total yield in the dry season but was approximately the same as the average response to the optimum nitrogen treatment.

In the dry areas of the country, Danckwerts and King (1984:26) show that stocking rate, expressed as grazing days per hectare, can vary from 87 grazing days to approximately 50 from one year to the next. This variation was caused by a variation in rainfall from 560mm to 420mm.

Since grassland productivity is the basis of all estimates of carrying capacity and therefore recommended stocking rates, the implications of this type of variability are obvious. Livestock production under extensive systems is essentially a stable operation over time, and is not amenable to major fluctuations from year to year in stock numbers. This being so, it begs the

Table 8.1: Influence of rainfall on yield of grassveld and response to nitrogen in the presence of phosphatic fertiliser (28 kg p/ha)

Season	*Rainfall mm*	*Control yield (kg DM/ha)*	*Maximum response to nitrogen kg*
1952/53	608	2188	2873
1953/54	756	5272	3456

Source: Booysen (1981:67)

question as to how precisely the accepted techniques for assessing veld condition and therefore carrying capacity can be brought to account in a system as variable as a natural grassland, without having an unacceptable effect on the animal system exploiting that grassland.

The erosion effect

Underlying the many references to degradation and the concern it causes is the fear of accelerated erosion resulting in soil loss greater than the natural regeneration rate: soil loss is an inherently irreversible process. This is essentially Sandford's condition of a decline in the productivity of land and water resources which is irreversible except at a prohibitive cost of recovery (Sandford, 1982a:44). Fear of this condition is either implicitly or explicitly stated in the bulk of the comment on the degradation issue. This being so there is a remarkable paucity of research evidence either supporting or contradicting the predictions of accelerated soil loss from degraded grazing land.

The problem was recognized as early as 1955 by Penz (1955:712) in his review of the state of erosion in South Africa. It is noteworthy however that he placed great emphasis on the effect of the encroachment by arable farming into grazing areas as a primary cause of denudation and soil erosion. Scott (1981:277) stresses the important role of both canopy and basal cover in dissipating the kinetic energy of falling raindrops, energy which would otherwise pound and pulverize the soil surface. He notes that cover is included as one of the factors in the SLEMSA (Anon: 1976) model used for predicting soil loss under various conditions. As may be expected, in this model soil loss is considered to be inversely proportionate to plant cover, which primarily determines energy interception. The relationship is a generalised one, relating soil loss to percentage of energy intercepted. Scott (1981:277) makes the link between veld management practices which will in the short and the long- term affect both canopy and basal cover, and the degree of erosion which will result from these practices. He notes nonetheless, that 'precise research data on soil loss from veld under South African conditions are, however, almost non-existent...'(ibid.).

One of the reasons for this is the cost in time and resources required to obtain data. Stocking (1985:7) reviews the available approaches to obtaining

Table 8.2: Soil loss from uncultivated land in Eastern Transvaal

Catchment	*Years*	*Sediment t/ha/annum*
Mpamo	17	2.96
Mgwempisi	2	5.00
Usutu	12	0.58
Komati (A)	19	0.18
Komati (B)	17	0.59

Source: Parsons (1986)

soil loss and run-off data and concludes that the ideal method, of field experiments using natural rainfall, would require data for a minimum of ten years to be valid. The cost of such programmes has lead to the widespread use of soil-loss estimating equations such as SLEMSA (Soil Loss Estimating Model for South Africa) and USLE (Universal Soil Loss Estimator). The latter was developed in the United States (Wischmeier and Smith, 1978) and has been standardised for use in many countries. While obviously having a valuable role as an aid in planning and administration (Crosby et al., 1983:8) USLE suffers from certain practical weaknesses arising from the necessity to establish values for the parameters used in the estimation. One such is the T-value, or soil loss tolerance which must obviously be related to the rate of natural soil genesis. In the United States this is estimated to be between 2.2 t/ha per annum and 11.2 t/ha per annum (Stocking, 1984:51). For South Africa values between 3 t/ha per annum and 9 t/ha per annum have been suggested (McPhee & Smithen, 1984; cited by Scotney & McPhee, 1990:6), for cultivated land. These estimates are determined largely subjectively, and all exceed accepted rates of soil formation. They appear to be used as a short-term solution to proceeding towards a more appropriate lower value (Scotney & McPhee, 1990:6).

Given these problematic issues and the very pressing issue of soil loss from cultivated land, it is to be expected that empirical evidence on soil loss from grassland is sparse. The problems are exacerbated by the fact that soil losses from the small plots used to calibrate the USLE cannot easily be related to measured soil loss from large catchments; a further problem is the reliance on highly generalised judgement values of cover and management factors when using the USLE in grassland or forest areas (McPhee et al., n.d.).

What evidence is available suggests that the rate of soil loss from grazing land may be well below levels which could cause concern. In a survey of five catchments feeding reservoirs in the Eastern Transvaal, Parsons (1986:1) found the rates of soil loss shown in Table 8.2. The high value for Mgwempisi was held to be anomalous because of the short period of accumulation. These losses compare well with the T-values for cultivated land quoted above, and Parsons (1986:2) doubts that the higher sediment yields are a consequence of overgrazing.

Table 8.3 Summary of soil losses reported in selected studies in the RSA (t/ha/annum)

Source	Soil losses under practices: Conventional		Conservation		Natural Veld	Slope %
Scott (1951)	2.0	(burn and graze)	0.6	(graze only)	0.09	8.00
Sherry (1959)	113.7	(brushpiles burnt)	4.6	(not burnt)	–	25.00
Haylett (1960)	61.9	(continuous crop)	37.4	(rotations)	0.7	7.00
	21.1	(continuous crop)	14.3	(rotations)	0.7	3.75
Du Plessis & Mostert (1965)	8.2	(continuous crop)	5.9	(rotations)	0.2	5.00

Source: Scotney & McPhee (1990:8)

Scotney and McPhee (1990:8) provide summarised data from a number of classical studies employing natural runoff plots (vide Stocking, 1984) and point out the accelerated losses associated with conventional management practices. These results are presented in Table 8.3.

A striking feature of these results is the contrast between the worst grassland and the results from cultivated land. Snyman (1985:116), in a study involving rainfall simulators, found a highly significant relationship between basal cover and soil-loss, and established the relationship shown in Table 8.4. Even the highest soil loss falls within at least one quoted range of acceptable T-factors (Stocking, 1984: 51).

Danckwerts (1990:15) in a trial testing the effect of particularly adverse burning and grazing practices adopted by farmers, recorded a loss of 0.223 t/ha from the most severe treatment. This was some 2% of the loss normally considered acceptable from cultivated land (10.0t/ha/annum) and far less than the rate of soil genesis. Danckwerts (1990:15) questions whether in fact soil loss is a major issue on grazing land.

Venter et al.(1989:89) have reported on a particularly relevant piece of work, on the boundary of KwaZulu and the Mfolozi Game Reserve. This study proceeded from the management problem of maintaining species diversity in a game reserve, together with the problem of long-term soil loss which, under

Table 8.4: Plant-succession/soil-loss relationship

Successional stage	*Mean soil loss (t/ha/annum)*
Climax	1.8
Sub-climax	2.3
Pioneer	4.7

Source: Smyman (1988:89)

specific systems, could exceed the rate of soil genesis. The researchers used a 'fenceline contrast study' comparing certain variables in KwaZulu and those in the Mfolozi Game Reserve. The KwaZulu area is described as having 'a long history of abusive grazing practices by domestic livestock, in particular cattle' (Venter et al, 1989:89). This is contrasted with the Mfolozi Game Reserve, where the stocking rate with large herbivores had been reduced to 60% of the recommended carrying capacity during the period from 1982-85. The basic assumption was that a fenceline study would indicate whether there was a higher long-term soil loss rate in KwaZulu as compared to Mfolozi; the study should also indicate whether there was a greater or lesser species diversity on either side of the fence.

The study area fell within the Tropical Bush and Savanna veld types (Acocks, 1975 cited by Venter et al., 1989:90). The measurement of short-term soil loss was by means of 48 Gerlach troughs (Morgan, 1979 cited by Venter et al., 1989:90) which were randomly located in pairs at 24 sites in the study area, 12 on each side of the fence. The Gerlach troughs were sited to capture run-off and soil loss from small catchment areas which could subsequently be measured by standard procedures. Long-term soil loss was measured by comparing the depth of the A horizon on either side of the fence. The assumption was that if the KwaZulu area had suffered accelerated soil loss over a long period, the A horizon depth would be significantly lower than that on the Mfolozi Game Reserve side. Since there was a clear demarcation between the A and B horizons, the measurement was simplified. The relationship between the degree of soil erosion and production potential was measured by means of harvesting the herbage from protected exclosure plots on either side of the fence. The comparison between grass species richness and diversity was measured by assembling grass species frequency data using quadrats randomly placed within each plot, and deriving an index of similarity to compare conditions on either side of the fence. The study was conducted over two rainy seasons.

The results indicated that over the two-year test period, short-term soil loss was consistently higher from the KwaZulu sites than from the Mfolozi Game Reserve sites, although no significant difference could be established. The long-term erosion levels, measured by means of the variation in A horizon depth, was again consistently higher on the KwaZulu side, but again no significant difference could be accounted for. There was no significant difference between KwaZulu and Mfolozi, in terms of herbage accumulated in the exclosure plots. As could be expected, when grazed there was a significant difference between the two areas. Plainly, this was due to the variation in grazing pressure applied to the two areas. While there were significant differences in species richness at the various positions on the slope, there was no significant difference between species variability at the pairs of plots measured in KwaZulu and Mfolozi Game Reserve. The authors conclude in their discussion that, for all the variables measured, the position of the plot on the slope made a greater contribution to variability than whether the plot was situated in KwaZulu or Mfolozi, and summarised that:

> ...the perceived management problem in following a non-interventionist policy therefore appears to be unfounded: both conceptually, when considered in the light of Natal Parks Board objectives for game reserves under their control and practically, when the herbage accumulation results are considered (Venter et al., 1989: 95).

They do add the cautionary note that it was assumed that the grazing pressures between the two areas had varied considerably over the years. Since the underlying assumption leading to the research was that, in fact, the KwaZulu area had been subject to a long history of abusive grazing by domestic livestock, this is a somewhat unexpected cautionary statement at the end of the paper. What can be concluded from this work is that, whatever the history of grazing conditions on either side of the fence, using a number of important degradation indicators as parameters, the conditions in KwaZulu are no worse than those in a well-protected and presumably well-managed conservation area.

Conclusion

The information presented in this chapter has not been intended to challenge or directly criticise the quality of the research carried out either on grassland dynamics or on erosion. In addition, it does not purport to be a comprehensive review of such research. It has been provided to show that, even under controlled research conditions, the assumed relationships upon which the present understanding of grassland dynamics are based, are not sufficiently consistent to be used as the basis for policy prescriptions. This applies in the case of a grassland resource such as that employed by Zulu cattle owners. The effect of applying the research on commercial ranches might mean no more than a variation in gross margin for a farmer who, since he is proceeding with the recommendation, could presumably afford the cost of an error. In the case where the recommendation amounts to social experimentation on a grand scale, resulting in fundamental changes to the economic and social fabric of an entire people, it is submitted that the information is not strong enough to justify so drastic an action.

In the past it has been tacitly assumed that the technical or ecological arguments for destocking were so overwhelming that they rendered examination of the economic aspects unnecessary. It is argued here that, in fact, the technical evidence is so fragile it does not of itself present a valid case for destocking. This being so, if a case is to be built, it can only be based on analysing the economic effects of destocking. A superficial calculation was provided here which suggested that there could be economic loss as a consequence of destocking. In a society in which cattle are so intricately bound into the social and material welfare of the communities which own them, far deeper analysis and careful calculations of welfare gains and losses will be required before action as drastic as destocking can be proposed.

9
An Analysis of Feed Demand and Supply for Pastoral Livestock: The Gourma Region of Mali

Peder N. de Leeuw, Lassine Diarra and Pierre Hiernaux

Introduction

The Gourma is a typical Sahelian rangeland with a sparse herbaceous cover dominated by annual species. Herbage production is mainly determined by soil moisture availability, which in turn results from the interplay between rainfall and landscape characteristics (Penning de Vries and Djitèye, 1982; Hiernaux et al., 1990). These interactions cause intricate patterns of run-off and run-on, thereby increasing resource heterogeneity and diversity in space and over time.

Rainfall in the Gourma is extremely variable in space and over time, increasing from 100 mm in the north to 400-500 mm in the south. As a consequence, herbage growth is limited by soil moisture below 250 mm and increasingly by soil nutrients further south (see de Ridder and Breman, in this book). In essence, the Gourma is an abiotically controlled ecosystem; changes in herbage productivity and species composition occur regardless of the level of herbivory (Ellis and Swift, 1988). Although herbage removal usually reduces seasonal yield, as long as flowering and seed production of the component species are not impeded, the long-term effects of herbivory are minor or restricted to sacrifice areas around waterpoints and settlements (Hiernaux et al., 1990).

The Gourma is mostly utilised by nomadic and transhumant pastoralists of great tribal diversity, who show the 'recurrent features of pastoral opportunism: i.e. fluctuating stocking rates and migratory patterns of forage exploitation' (Behnke and Scoones, 1991:24).

Being part of the Sahelian zone, there has been great concern about continuing environmental degradation for which overgrazing by livestock is held partially responsible, based on the view that traditional livestock systems are inherently destructive to the environment. It is also believed that development (such as better veterinary services and increased water supplies) coupled with reduced pastoral mobility has exacerbated the degrading process (Lamprey, 1983; Mabbut, 1985).

In contrast to alarmist predictions of an environmental catastrophe, others (Sandford, 1983a; Ellis, 1991; Winrock, 1991) argue that there is inadequate evidence that livestock and their owners are to blame for degradation. Instead, many developers and scientists have failed to understand the ecological and social components of traditional production systems, as well as the survival strategies and production goals of pastoralists; this failure is often cited as the reason for both the poor performance of development projects and the unqualified condemnation of pastoralists as destroyers of their own environment (de Haan, 1990).

In this chapter, it is argued that the application of scientific knowledge is an essential prerequisite for sustainable resource management. Scientific inputs have the following components when applied to a specific region:

(i) a thorough analysis of the resource base, with emphasis on feed and water supplies and demands in the short- and the long-term

(ii) the establishment of data banks on livestock systems, comprising utilisation strategies, mobility patterns, offtake and marketing potential, etc.

Our emphasis will be on the utility of assessing feed supply and demand in abiotically controlled ecosystems. In this context, the chapter is a contribution to the ongoing debate and the critical reviews of the carrying capacity concept brought about by de Leeuw and Tothill (1990), Behnke and Scoones (1991) and Bartels et al. (1990) - see also their chapters in this book. The Gourma region is considered an appropriate example, as it has been subject to intensive research extending from the early 1970s (Gallais, 1975; Boudet et al., 1971) to the late 1980s (IEMVT, 1989; Hiernaux et al., 1990).

We first consider the demand for feed supplies, by analysing the temporal trends in livestock numbers that have used the Gourma as a grazing area. This is followed by an assessment of the entire region's feed supplies, making use of data sources emanating from research by the International Livestock Centre for Africa (ILCA) on the vegetation dynamics and trends. Three major aspects of the feed supply will be considered: (i) the inter-annual variation of forage production in space and over time; (ii) the impact of herbage disappearance during the dry season and (iii) the effect of the spatial distribution of water supplies on forage accessibility.

Livestock population trends

Livestock population trends in the Gourma cannot be analysed in isolation, since this area forms part of a much larger pastoral production zone characterised by large-scale mobility of stock and people. These mobility patterns have been recorded in detail over the last two decades and have been recently summarised and mapped (IEMVT, 1989; Cisse and Gosseye, 1990; van Duivenbooden and Gosseye, 1990). Livestock populations of the Gourma are influenced by the region's spatial characteristics, which together with the seasonal variation of water and grazing resources, affect utilisation patterns in the short- and long-term. Four major features determine these patterns (see also Fig. 9.1).

(i) To the North and East the Gourma is bounded by the river Niger, which provides a major source of permanent water. Along the flood plain there is a dense sedentary population, dependent on irrigated cropping and

Figure 9.1: Map of the Gourma and surrounding region of the Sahel in Mali

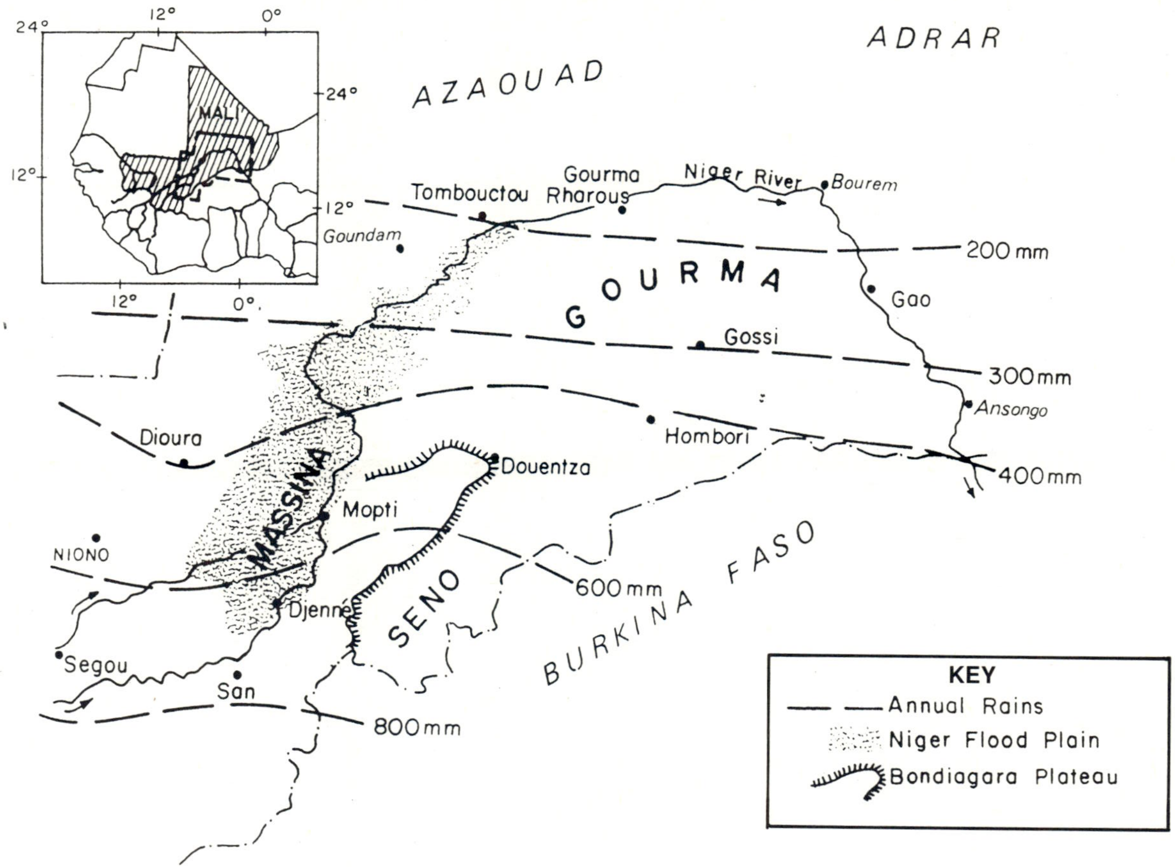

livestock husbandry, while in the dry season the river attracts many herds and flocks from the surrounding hinterland.

(ii) To the North-west the region is bounded by the lacustrine zone, which forms the North-eastern extension of the Niger River inner delta. Not only does this zone contain extensive dry season grazing resources, but it also supports a large sedentary population mainly engaged in irrigated cropping.[1]

(iii) Farther to the West lies the inner Niger Delta, which in addition to extensive rainfed and irrigated farming, has a large dry season livestock population.[2]

(iv) Towards the South-west of the Gourma, there are densely-populated areas (Bandiagara plateau, Koro), supporting mixed farming systems, with 0.2-0.3 million cattle and 0.4 million smallstock.

Because of its limited permanent water resources, the Gourma functions mostly as a crucial wet season area for herds and flocks, which rely on permanent water resources outside the Gourma during the dry season. Hence, the major axes of mobility are outwards with the onset of the dry season and inwards during the rains. The movements into the Gourma are prompted by the gradual flooding of the floodplains and lacustrine areas, resulting in scarcity of accessible grazing lands within the densely farmed areas. The Gourma is favoured as a wet season grazing area because rainfed farming is rare within the region (except in the South-west) and because water supplies are relatively abundant at that season, at least in years of average rainfall (Bourn and Wint, 1985b; Hiernaux and Cissé, 1983; Hesse et al., 1984).

Since the early 1970s, there has been a steady decline in the cattle population of the Malian Sahel. In 1967 about 83% of the 5.2 million cattle were found in areas receiving less than 600 mm of annual rainfall. During the 1972-4 drought the total Sahel herd dropped from 4.7 to 2.6 million. Thereafter up to 1982, there was a slow recovery to 4.0 million, followed by heavy losses during the 1982-85 drought when a new low of 2.7 cattle million was recorded (IEMVT, 1989).

The same trend was evident in the North-east of Mali, where an overall decline of 44% occurred between 1967 and 1984 (Table 9.1). The rates of cattle

---

1. In 1979 the lacustrine area had a population of 212,000 people, 130,000 cattle, 600,000 smallstock. There was a total of 18,000 ha irrigated land (PIRT, 1983).

2. The livestock population within the Delta itself was estimated at 0.8 million cattle and 1.5 million smallstock in 1979 (PIRT, 1983). In 1984, the cattle population was estimated as 1.0 million with 0.7 million smallstock (IEMVT, 1989). During the dry season in 1981, the floodplain region harboured 1.2 million cattle and 0.5 million smallstock (de Leeuw and Milligan, 1983).

Table 9.1: Long-term trends in the cattle population of the Gourma and its surrounding hinterland (in '000 head)

	Year			
Area[a]	*1967*	*1979*	*1984*	*% change*
North	355	92	69	(-81%)
East	468	129	132	(-72%)
Central	395	206	206	(-38%)
North West	235	130	147	(-37%)
South West	313	453	394	(+26%)
	-----	-----	-----	--------
Total	1,766	1,010	989	(-44%)

Note: [a]Location of cercles: North: Timbouctou; Bourem.
East: Gao; Ansongo.
Central: Gourma; Rharous; Bandiagara.
North West: Dire; Goundam.
South West: Douentza; Koro.

Source: PIRT (1983); IEMVT (1989)

population decline increased with aridity, from 20-40% in the 400-500 mm zone to 70-80% in the <200 mm zone. It is noted that livestock numbers increased in the South-west, probably due to the combined effects of livestock being shifted further south and increased livestock ownership by sedentary farmers. Although the arid hinterland lost 75% of its cattle, the Gourma as an entity was less affected and retained its position as an important cattle area for the entire North-east part of the Malian Sahel. Based on detailed studies of many tribal groups of pastoralists in the early 1970s, Gallais (1975) estimated that the maximum number of cattle using the Gourma amounted to 1.2 million, together with 2 million smallstock and 30,000 camels. He further emphasised that many herds spent only limited periods in the Gourma; for instance, the 300,000 head coming from the West remained for only two months during the rainy and early dry season. Boudet et al. (1971) estimated the number of cattle entering the Gourma from the North and East at 0.88 million or about 70% of the total cattle population in the North, East and Central areas (Table 9.1). Whatever the reliability of these census data, there is little doubt that prior to the 1972-4 drought the Gourma was an important grazing area for the entire North-eastern part of Mali.

The decline of livestock populations in the Gourma was confirmed by a series of aerial reconnaissance flights carried out in January and March 1983 and again in late August-early September 1984 (Milligan, 1983b; Bourn and Wint, 1985b). The overall cattle population amounted to 0.36-0.40 million in the

Table 9.2: Populations of cattle and smallstock in the Gourma during early 1983 and late 1984 by ecological zone ('000 head)

	Cattle			*Smallstock*		
Zone	*Jan 83*	*Mar 83*	*Sept 84*	*Jan 83*	*Mar 83*	*Sept 84*
North Sahel	178	114	101	233	193	374
South Sahel	113	144	215	187	232	457
Sahelo-Soudan	107	99	168	163	156	307
	----	----	----	----	----	-----
Total	398	357	484	583	581	1138

Sources: Milligan (1983b); Bourn and Wint (1985a)

dry season in 1983 and increased to 0.47 million in the latter part of the 1984 wet season (Table 9.2).The rainy season estimates are probably comparable to those of Gallais (1975) and Boudet et al. (1971); they represent the maximum population of cattle which tends to occur towards the end of the rainy season. If in 1984 the potential hinterland population, together with the 'autochthonous' Gourma herds are estimated at 1.0 m (Table 9.1), then the wet season population, as counted in 1984, represents only 50% of the potential influx. The 1984 rainfall was very low, particularly in the North, resulting in a continuous belt devoid of vegetative cover all along the northern boundary and representing at least 20% of the total Gourma area (Bourn and Wint, 1985b; Hiernaux and Justice 1986). Although this lack of rain may have reduced the entry of livestock into the northern fringe of the Gourma, it may also reflect the declining trend of livestock populations in the surrounding hinterland as shown in Table 9.1.

Monitorings of spatial distributions of livestock populations intensified in the late 1980s and are currently being analysed. The outcome will confirm whether the downward trends are continuing, or conversely that utilisation patterns follow opportunistic adaptations to short-term changes in resource availability (Dr. A. Diallo, pers. comm.).

Feed resources of the Gourma

The distribution and total quantity of feed resources in the Gourma region were assessed by taking into account two main variables: mean annual rainfall and landscapes. Given the North-South rainfall gradient, the Gourma was divided into three main ecological zones, the boundaries of which follow latitude: the northern Sahel with an average rainfall below 250 mm, the southern Sahel with 250-400 mm and the Sahel-Soudan zone receiving over 400 mm of annual rainfall (Hiernaux and Cissé, 1983; Hiernaux and Justice, 1986).

Landscapes fall into four major groups, shown in Table 9.3. The main forms, of Group 1, occupy 50% of the Gourma and are more common in the north than in the south. Group 2 occupy about one third of the total area, and are more important in the south. Group 3 and Group 4 cover 11% and 5%, respectively,

Table 9.3: Distribution of major landforms in the Gourma region in the three major ecological zones

	Northern Sahel	*Southern Sahel*	*Sahel-Soudan*	*Total*
Land forms		*in % of land area*		
Group 1				
Sand dunes	27	25	2	19
Sandy peneplains	20	5	–	9
Sandy cover over bedrock and laterite cuirasses	13	20	37	22
Group 2				
Cuirasses, rock outcrops	24	34	46	34
Group 3				
Alluvial plains	5	13	15	11
Group 4				
Floodplains	11	3	–	5

Source: Boudet et al. (1971); IEMVT (1989)

of the total area.

Given the large variability of rainfall, estimating average feed resources is a risky task. Nevertheless, based on previous surveys and current ILCA research, average yields for each landform can be established, which together with their variation across time, makes it possible to estimate (i) the mean contribution of each landform to feed resources and (ii) the changes in contribution across years.

In terms of average feed resources, Group 1 landforms with sandy soils provide about 55% of the total from half of the total land area. This average was based on weighted means of herbage yields of 0.4, 0.7 and 1.0 t DM/ha for the Northern Sahel, Southern Sahel and Sahel-Soudan zone respectively, and were derived from yield data recorded over a 7-year period as discussed below. In contrast, Group 2 landforms with shallow stony soils are less productive, contributing only 14% of feed supplies from a land area more than twice that percentage (Table 9.4). The Group 4 Floodplains are highly productive, with 2 t DM/ha of usable forage. Average, lower and upper limits of herbage yields for each landform have been arbitrarily estimated to predict the likely fluctuations in feed supplies across years (Tables 9.4 and 9.5). In an average year, total feed supplies reached about 5 million t of dry forage or 0.6 t DM/ha. In years with low and high rainfall, supplies were approximately

Table 9.4: Average, low and high estimates of feed supplies of major landforms in the Gourma region (in million t)

	Landforms				
	Group 1 Dunes, sandy soils	*Group 2 Cuirasses, gravels, rocks*	*Group 3 Clay plains, valleys*	*Group 4 Flood-plains*	*Total supply*
Average: t DM/ha	0.65	0.25	0.80	2.00	0.60
Total supply, 10^6t	2.87	0.73	0.75	0.85	5.20
% of total	55	14	15	1	100
Low: t DM/ha	0.25	0.10	0.60	2.00	0.32
Total supply, 10^6t	1.05	0.28	0.57	0.85	2.75
% of total	40	10	21	31	100
High: t DM/ha	1.10	0.50	1.20	2.00	0.94
Total supply, 10^6t	4.81	1.49	1.03	0.85	8.18
% of total	59	18	13	10	100
% of total land area	50	35	10	5	100

50% below or above the average. Owing to the greater response to better rainfall, Group 1 sandy sites contribute 60% of the total supplies in good years as compared to 40% in poor years. Conversely, in years with low rainfall, about half the potential feed was generated by sites receiving extra moisture supplies either from run-on water or from flooding, indicating the potential risks of over-exploitation of these feed resources in below-average rainfall years. Yield responses to soil moisture conditions across years are required in order to assess the probabilities of attainable yield levels for vegetation types associated with the four major landforms. While average contributions to the overall feed supply constitute an essential bench-mark for rating the productivity of landscapes, the degree of response to variable rainfall conditions ultimately determines the amplitude of the overall supply across years. Consequently, relationships between yield and rainfall will first be examined, and secondly, changes in floristic composition across space and time will be analysed. In the latter case, emphasis is on grass/herb ratios, since the grass content in the total herbage is an important factor for the feed supply in areas where close to 80% of the livestock mass consists of cattle (Table 9.2).

Inter-annual variations in feed supply

Relationships between annual rainfall and maximum herbaceous biomass were analysed over the period 1984-90. For this purpose, the rain use efficiency

Table 9.5: Weighted average, low and high herbage yields (t DM/ha) in the Northern and Southern Sahel and Sahel-Soudan zones

Zone	*Average*	*Low*	*High*
Northern	0.50	0.15	0.85
Southern	0.60	0.30	0.95
Sahel-Soudan	0.70	0.35	0.95
	-----	-----	-----
Total area	0.60	0.32	0.94

factor (RUE) was introduced; this is the amount of herbaceous biomass in kg DM/ha-1 per mm of rain received. This factor is appropriate for herbaceous vegetation that consists mainly of annual species, as it captures the differences in yield due to variable rainfall distribution for the same annual amount of rain (Le Houérou et al., 1988).

The analysis was restricted to sites on sand dunes and other landforms with sandy topsoils (i.e. Group 1 landforms). On most sandy soils, infiltration rates are high and run-off is low; hence, responses to rainfall are directly related to plant establishment and growth (Hiernaux, 1984). In total, 79 paired rainfall and maximum biomass data from 13 sites were used. As these sites were located along a 250-km long North-South transect following the rainfall gradient, a sub-division between north and south was justified, resulting in 40 and 39 data pairs respectively.

It is noted that, as shown by Le Houérou et al. (1988), herbage yield is more variable than annual rainfall, while rain use efficiency has similar coefficients of variation as those of yield (Table 9.6). This indicates that annual rainfall alone may be a weak predictor of herbage productivity, despite the significant regressions of annual rainfall on yield[3].

During the seven-year analysis period, 1984 was the worst year, while the highest herbage yields were recorded in 1988 and 1989 across the entire transect (Table 9.7). The other four years were intermediate; within this group, 1990 ranked best in the north and 1987 in the south. It is noted that annual rainfall for all but the best years was similar; 107-120 mm in the North and 165-210 mm in the South. On the whole, the yields in four out of the seven years are reasonably close to the averages for sandy soils used in the estimations of total feed supply, while values for 1984-5 and 1988-9 are similar to the lower and upper limits shown in Table 9.4.

Vegetation communities on loamy sites are mostly confined to areas with shallow soils, where usually surface sealing and compaction impede water

3. North : Y = -402 + 7,05 X (N = 40 : R^2 = 0.52)
   South : Y = 273 + 5,26 X (N = 39 : R^2 = 0.35)

   Y = kg DM/ha; X = mean annual rainfall in mm

Table 9.6: Mean annual rainfall, maximum herbaceous biomass and rain utilisation efficiency (RUE) on sandy soils, 1984-90

	North		*South*	
Mean rainfall, mm (CV%)	144	(37)	205	(28)
Mean maximum biomass, kg DM ha-1 (CV %)	615	(86)	804	(64)
RUE, kg DM ha-1 mm-1 (CV%)	3.70	(76)	3.76	(49)

infiltration and conditions for plant germination and growth are poor. Only four sites have been regularly monitored since 1984. Out of a total of 24 observations, end-of-season yields of below 100 kg DM/ha comprised 60%; of the remainder, 25% produced 100-200 kg, while maximum yields never exceeded 500 kg DM/ha. Thus, the contribution of these sites to feed resources is low but fairly stable over time.

Although plains developed from colluvial and alluvial deposits cover some 11% of the area, only a small portion is highly productive, mostly associated with valley-bottoms and depressions benefitting from run-on moisture. Away from the low-lying areas, slopes (and therefore run-off) increase which, coupled with compacted surfaces, produce bare land with scattered pockets of low herbage cover where growing conditions are slightly more favourable. However, range types in ungrazed depressions can produce up to 5 t DM/ha. Grass cover consists of *fonio*-type grasses (*Panicum leatum, Echinochloa* spp.) that are harvested for grain and therefore protected (Maiga et al., 1991). When freely accessible, end-of-season yields are much lower owing to heavy grazing with wet-season stocking rates of up to 2 TLU/ha. These sites, although limited in extent (2-3% of the total land area) constitute a key resource for grazing (Bouba-Kaou, 1990) and mineral supplementation of livestock ('cure salée', see Soumare, 1986) as well as for the food security of the local population. Therefore, balanced use and proper

Table 9.7: Rainfall and herbage yield on sandy soils, 1984-90

	North		*South*	
Year	*Rainfall mm*	*Yield t DM/ha*	*Rainfall mm*	*Yield t DM/ha*
1984	116	0.09	199	0.23
1985	119	0.29	184	0.68
1986	107	0.34	180	0.64
1987	118	0.33	165	0.81
1988	203	1.20	284	1.68
1989	193	0.96	206	0.90
1990	120	0.81	210	0.47

management are essential tools to sustain their productivity (Maiga et al., 1991).

Herbage supplies for livestock are also influenced by the evolution of the floristic composition. Although most livestock species, including cattle, are known to become less selective when feed shortages are acute, the dietary preference of cattle remains grass herbage. Hence, the grass content of the total herbage is an important component when estimating total feed supplies (Fig. 9.2). The effect of the low yield levels in 1984 not only drastically reduced short-term feed supplies, but also led to a general fall in species diversity, as many species, in particular longer-cycle grasses, failed to reach maturity and to set seed. This decline was exacerbated by the subsequent series of dry years (1985-7) which further reduced the total number of species and the ratio between grasses and dicots. Dicots are often pioneer species and during these dry years, dominated on many sites. Only during the good rainfall year of 1988 did grasses begin to gain dominance in most rangeland types on sandy soils. While from 1985-7 to 1988-9 average total herbage yields doubled or tripled (Table 9.7), the rise of grass herbage yield was even steeper, owing to the rising grass content in the total herbage (Fig. 9.2).

Both the greater species diversity and a larger proportion of grasses in swards increased yield stability over time; annual grasses respond more rapidly than dicots to better soil moisture conditions due to their tillering ability, the flexibility of their germination strategies and their growing cycle duration.

Biomass disappearance rates

Calculations of feed supplies based solely on end-of-season herbage yield fail to take into account the disappearance of biomass during the nine months of the dry season, due to causes other than the removal by livestock. These rates were regularly obtained on several permanent sites in the Gourma (Hiernaux, 1989). Biomass decline during the dry season occurs in three phases: a rapid decline early in the season due to fall of leaves, fruits and/or seeds; a slower rate during the cool and dry months, followed by an increasing rate when temperature and humidity rise in March/April, culminating in rapid decomposition after the first rains in May-June. Since in many of the permanent sites, biomass was recorded throughout the growing season often up to the end of October, disappearance rates from the first phase were quantified. Residual biomass varied from 10 to 93% of maximum biomass, with 62% of the sites retaining over 60% (Fig. 9.3). During the second phase decline is slower but variable, and rates during the first and second phase do not appear correlated, although Hiernaux (1989) showed strong links between dry season grazing pressure and disappearance rates.

In view of the large variation in disappearance rates, it is not clear how dry-season feed supplies should be adjusted. Perhaps a biological model that incorporates the principal causes of decline should be developed, as proposed by Hiernaux (1989). Provisionally it appears that the average rate of decline is about 60%, which would translate into a composite rate of 9% per month. This rate is similar to a total disappearance rate of 65% in the rangelands of Central

Figure 9.2: Contribution of grasses and dicotyledons to total herbage in the Northern and Southern Gourma, 1985-1990

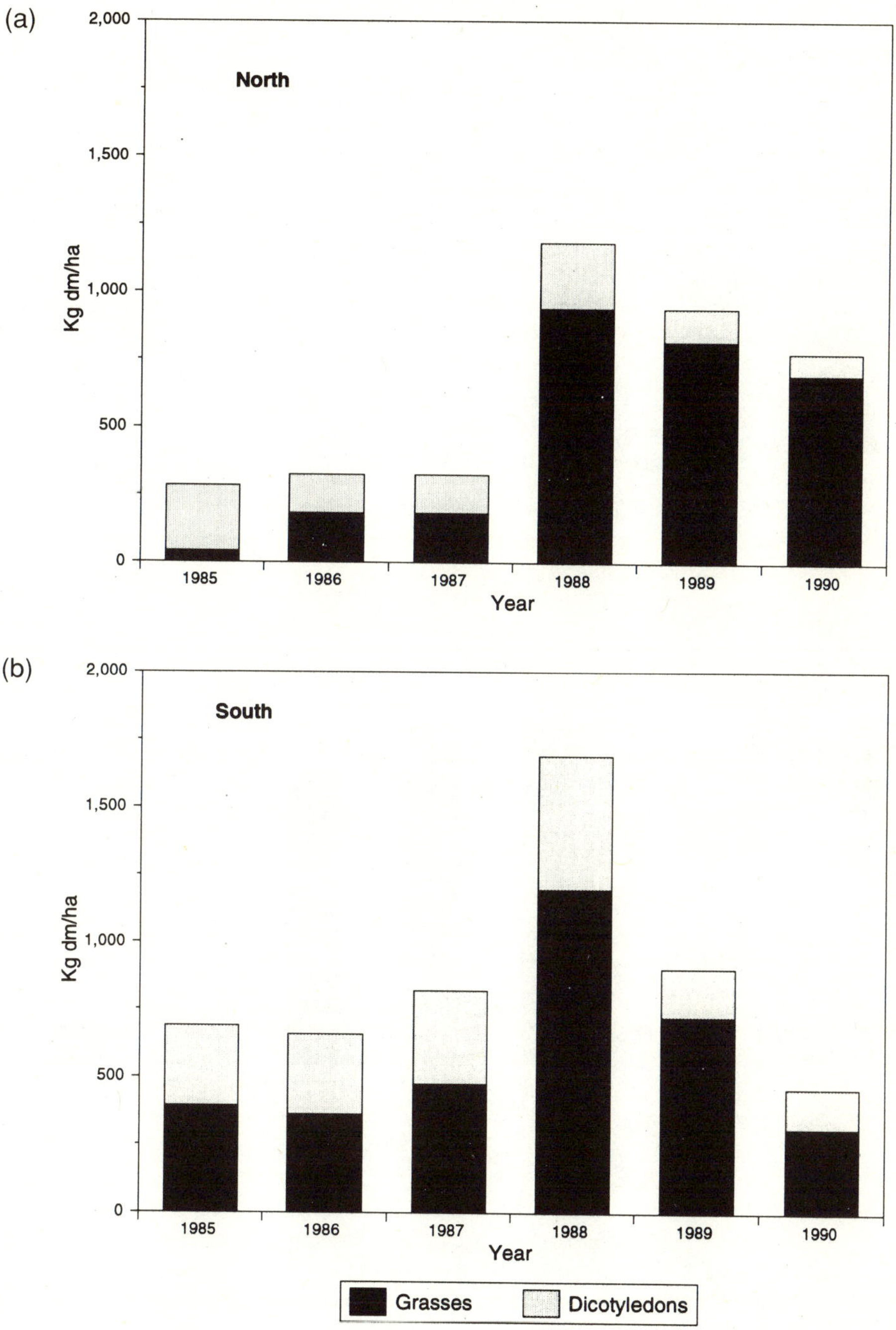

Mali in the 300-500 mm rainfall zone (Diarra, 1983), but much higher than the monthly rate of 4% per dry month (converting to a cumulative decline of 26%) reported by Wylie et al (1988) in the pastoral zone in Niger.

Distribution of water points

The Gourma is endowed with a large diversity of water sources, which vary in terms of the quantity, quality and duration of available water and the ease of extraction. The Niger River functions as a important water source and attracted some 100,000 cattle during the 1983 dry season or 30% of the total population (see Table 9.2). Other permanent sources are the large lakes near Gossi and Tessi, while numerous permanent wells are located along the piedmonts of the 'buttes' in the South-west (Hesse et al., 1984). Additional sources are the deep wells and boreholes constructed by the government or by aid projects (Milligan, 1983; Hiernaux and Cissé, 1983). Temporary wells are

Figure 9.3: Frequency distribution of monthly disappearance rates of herbage biomass, for the first month and following 6 to 8 months of the dry season (Gourma, Mali)

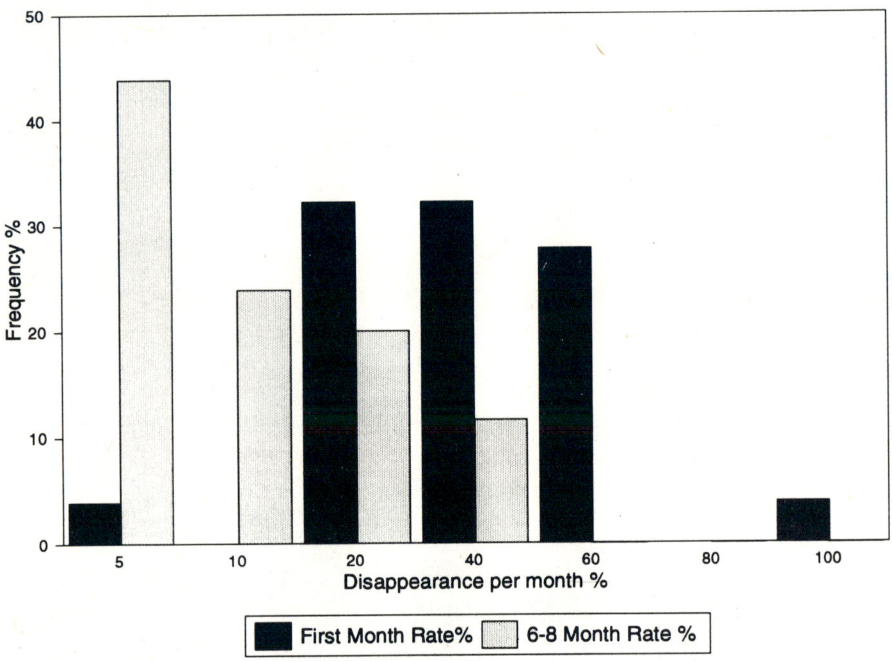

Table 9.8: Proportion of grazing land (< 7.5 km, 7.5-15.0 km and >15 km) from water for the three Gourma ecological zones (in % of land area)

	Distance (km)		
Zone	*< 7.5*	*7.5 - 15.0*	*>15.0*
North Sahel	47	19	34
South Sahel	37	21	42
Sahel-Soudan	55	26	19
	---	---	---
Mean for Gourma	44	22	34

Source: Adapted from Hiernaux and Cissé (1983, Fig. 1 and Table 2)

dug along river beds and depressions to exploit water tables at shallow depth. During the rains, there are numerous pools in depressions and valleys, which provide water then and in the early dry season. Many of these not only provide stock water but are also the major locations for the 'cure salée' (salt cure) to remedy mineral deficiencies in livestock. More than 100 locations of this 'cure salée' are known to exist and a detailed account of when they are used, by which pastoral groups and for what type of cure has been published (Soumare, 1986).

Hiernaux and Cissé (1983) classified all known waterpoints in respect of their water capacity, duration of flow and ease of access. From this survey an index of water distribution for the entire Gourma was developed as a predictor of potential grazing pressure. The proportion of land in relation to distance to water was estimated for the three ecological zones (Table 9.8). This grouping of distance to water was based on the hypothesis that grazing orbits relying on one-day watering are usually confined within 7.5 km of a water point while, when stock are watered every second day, the distance from a water point may increase to 15 km. Beyond the 15 km limit, range utilisation was expected to be low, owing to inadequate access to water for livestock.

Climate and geomorphology of the area interact: areas close to the Niger River have a high index, whereas in the remainder of the Northern Sahel half the land lies outside the 15 km limit, in contrast to only 20% outside the limit in the South-west. While dry season water supplies are fairly stable, as they depend on permanent sources, wet season sources are much more rainfall-dependent; this leads to rapid contraction and expansion of potential wet-season grazing capacity. For instance, 1984 was considered a drought year affecting grass cover as well as water distribution (Hiernaux et al., 1990). During the 1984 aerial survey, open water surfaces (in pools, river beds etc.) were recorded for each grid of 80 km^2 (Bourn and Wint, 1985b). In the North Sahel only 24% of the grids showed visible water, out of which half were within 15 km from the Niger River, rendering the rest of the area nearly

Figure 9.4: Frequency distribution of herbage biomass late in dry season (April 1983) within and over 15 km of the water points (Gourma, 2,500 observations)

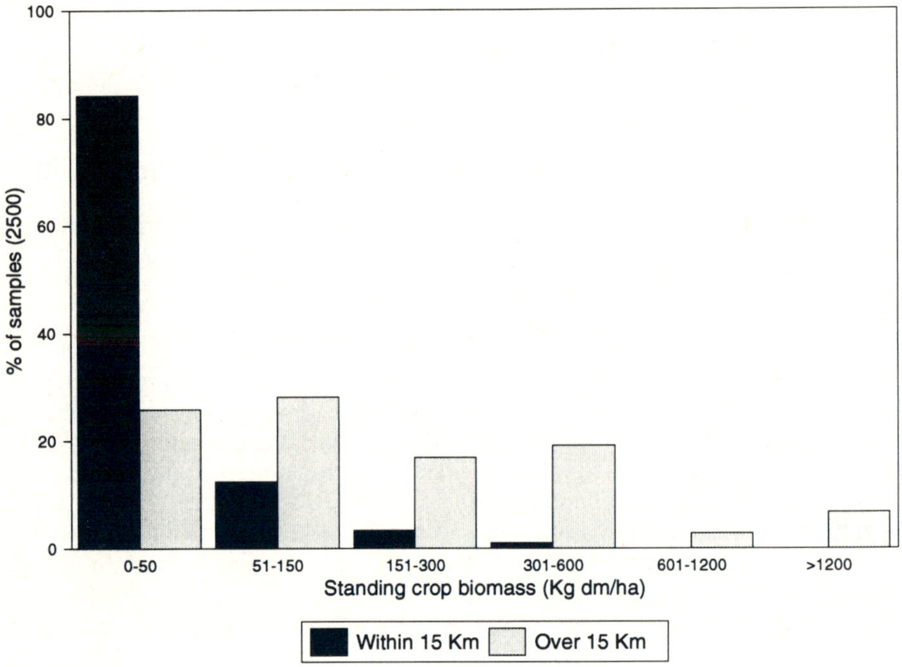

waterless. This may explain why only 25% of the total number of TLU was found in the North (Table 9.2). In the South Sahel and the Sahel Soudan zones, surface water was observed in 66% and 82% of the grids respectively.

Direct evidence of differential grazing pressure around waterpoints during the dry season of 1982-3 was reported by Hiernaux and Cissé (1983). Some 2,500 samples of biomass in 1 m^2 quadrats were taken in late April 1983 all over the Gourma. Three hundred samples were taken in areas over 15 km away from waterpoints, and 2,200 within 0-15 km distance. The frequency distributions of biomass in the two sample sets is shown in Fig. 9.4, indicating that, away from water supplies, about half the land had more than 150 kg/ha of biomass as compared to 5% in areas closer to water; mean yields were 360 and 45 kg of dry biomass respectively. The overall mean was 83 kg/ha, a clear indication of the precarious feed situation towards the end of the dry season in 1983.

Discussion

Feed Supplies

The analysis of the feed supply situation in the Gourma has been reviewed on the basis of existing information. The starting point for this analysis was that good long-term data sets existed of the yield potential of the Gourma rangelands and their temporal and spatial variability. Although the limitations of using relationships between annual rainfall and herbage yield are recognised, the value of rainfall for prediction of feed supply levels has been proven. Further precision of feed supply estimates can be achieved when habitat characteristics (landscapes, landforms, soils) are known and mapped as is the case for the Gourma (cf. Boudet et al., 1971; PIRT 1983; IEMVT, 1989).

In addition to inter-annual variability, the feed supply must be adjusted by the rate of dry season disappearance, from natural and grazing pressure causes, and to limitations of access due to the distribution of waterpoints. These adjustments are required when estimating dry season supplies on a monthly basis. As the end-of- season yields are the point of departure, further downward adjustment may be required for wet season grazing pressure. Given the continuous monitoring systems established by the Livestock Services Department and non-governmental organisations, mobility patterns and associated livestock densities are known, from which crude estimates of grazing impact can be made. These monitoring systems also provide information on the likely profiles of feed demand over time and in space.

Proper use factor

While feed demand can be calculated from known daily intake, a permissible consumption rate (i.e. the proper use factor) is required to determine the livestock carrying capacity for any given feed supply. These rates are usually fixed at fractions of 30-45% of the total herbage biomass (de Leeuw and Tothill, 1990). 'Proper use' implies that long-term productivity is maintained (Bartels et al.,1990) and that improper use or consumption rates exceeding the permissible limit will decrease forage production in subsequent seasons. These assumptions are appropriate for rangelands with perennial grasses, where negative feedbacks with high stocking rates are expected (Bartels et al., 1990; Belsky, 1986). In rangelands with annual grasses, herbage removal during the growing season will reduce dry season supplies, but there is little impact on subsequent seasons, provided seed production has not been impaired. Risk of improper use is further reduced as carry-over seed stocks are usually sufficient to allow establishment in the two following years. Therefore, in the Gourma rangelands, high removal rates are permissible without risking a decline in forage productivity. Some caution is required, however, as high grazing pressure may cause severe soil compaction due to trampling, which, combined with total removal of plant cover, may lead to wind and water erosion. Therefore, Wylie et al. (1988) recommended utilisation rates that would retain, at the end of the dry season, a protective cover equivalent to 0.2 t DM/ha.

Carrying Capacity

If the annual utilisation rate is set at 80% of the total end-of-season feed supply and reduced to 40% to allow for natural disappearance during the dry season and lack of access to stock water, the entire Gourma could theoretically support 570,000 TLU on a year-round basis in average years, and 900,000 or 300,000 TLU in good or poor years respectively. These numbers would be equivalent to region-wide stocking rates of 6.6, 10.4 and 3.5 TLU/km^2. In fact, the livestock densities recorded in the 1983 dry season and in the 1984 wet season varied from 3.5-5.0 TLU/km^2 in the northern part of the Gourma to 4.6-7.6 TLU/km^2 in the southern part, and are therefore well in balance with the predicted feed supplies. Using the methodologies described by de Ridder and Breman (1990), Veeneklaas et al. (1990:155, Table 6.26) calculated similar permissible densities in the order of 8-12 TLU/km^2 for the Gourma region.

Remote sensing

The continuous monitoring of feed supplies in pastoral areas is costly in terms of the manpower and transport required. Thus, although the need for regular surveys of range, water and livestock resources is recognised, existing monitoring systems in Sahelian countries usually depend on foreign aid to sustain operations. However, during the 1980s, the integration of satellite data into pastoral monitoring systems has developed rapidly. Early work focused on the use of vegetation indices as a means of measuring rangeland primary production (see Hiernaux and Justice 1986, for review). More recently, other applications to which satellite data can contribute include estimation of decadal or monthly rainfall, evapotranspiration, crop production and the extent of fires (Prince et al., 1990; Prince and Justice, 1991). Rapid developments in processing and analysis of satellite data, combined with the increased use of geographic information systems as a tool for spatial data management, have greatly increased the utility of remotely sensed data and facilitated their integration with other data such as landscape classification, distribution of water supplies, and human and livestock populations.

Conclusion

Efficient low-cost monitoring systems are an essential tool for predicting changes in feed and water resources and the likely responses from pastoral herders, in terms of opportunistic exploitation strategies. Such forecasting is necessary for the sound management of resources in the Sahelian ecosystems; sound management is considered a prerequisite for sustained development and environmental protection. A thorough understanding of the ecosystem dynamics is required to achieve flexible management strategies that can readily be adapted to changing conditions. It is also necessary to understand the pastoralists who manage the livestock, since without their co-operation and participation, sound management remains a pipe dream.

10
How Long can High Stocking Densities be Sustained?

Yvan Biot

Introduction

The previous chapters have shown how misleading the notion of carrying capacity can be when deciding on appropriate management of rangeland. The stocking densities which are most suited to any one situation are conditioned primarily by the management objectives of the producers, and the socio-economic and ecological environment in which they operate. The animal densities pursued by ranchers in a capital-intensive economy are such as to produce high quality meat, and are substantially lower than those practised by subsistence pastoralists in a labour-intensive economy and a highly variable ecology. Thus the carrying capacity of a given tract of rangeland is not only a management-oriented but also a management-dependent concept.

The chapter by de Ridder and Bremen in this book has outlined a state of the art methodology for the estimation of optimum stocking densities, based on food supply and demand functions derived from the Sahelian study by Penning de Vries and Djitèye (1982). It is usual in carrying capacity studies, however, to add an additional consideration, i.e. one of 'sustainability' or 'minimum environmental impact'. There may, indeed, be a price to pay in the medium and long term for pursuing certain management objectives. In the medium term, high stocking densities are often accompanied by massive deaths during droughts, and in the long term there is a problem of livestock-accelerated land degradation. On the other hand, keeping animal numbers below a minimum density may send the vegetation towards a different, and less productive, climax.

High stocking densities have been shown to lead to accelerated soil erosion - see, for instance, Blackburn (1984) and Wight and Siddoway (1982) for the United States, and Johns et al. (1984) and Buckley (1985) for the Australian savannas. The effect is characterised by an increase in bulk density of the topsoil, a decrease in soil infiltrability, and an increase in run-off and erosion. The latter two processes may affect the future ability of the land to produce plant biomass from which livestock feed is derived.

This chapter outlines a methodology for evaluating the capacity of rangeland to sustain high stocking densities in the long term. The chapter by Abel in this book deals with medium-term projections of livestock production. The present chapter is restricted to a study of the sustainability of livestock production as affected solely by physical processes, and does not discuss the political, economic and institutional context required for sustainable agricultural production. The focus here does not imply a rank order of importance among these issues.

Land degradation

A definition

The management objectives and socio-economic conditions of producers must be taken into account, in addition to the feed supply and demand functions of livestock production, in evaluating the physical processes which may lead to their production strategies not being sustainable. Abel and Blaikie (1989) illustrate this point well in their definition of land degradation (see below, p.175).

Land degradation is considered as a decline in the capacity of land to produce. However, not all decline in productive capacity is interpreted as land degradation: some decline occurs as a response to climatic events, and some decline can easily be remedied by management. In other words, production declines are only interpreted as land degradation when they are 'irreversible', defined in economic terms. The definition is further restricted by limiting degradation to processes which take place on a time-scale relevant to humans, rather than an indefinite one.

An important deduction is therefore that *not all processes of environmental change are necessarily processes of land degradation.*

Natural or human-induced ?

The concept of land degradation used here is not restricted to the consequences of management. It includes, for instance, soil erosion in semi-arid landscapes which is naturally faster than soil formation, and will invariably reduce soil depth and soil productivity in the long run, *irrespective* of whether there are animals on the range or not.

The distinction between natural and human-induced land degradation has traditionally been an important one for the natural resource scientist. It is doubtful, however, whether this distinction is an important one for the land user[1], where what matters is probably more adequately reflected in the following questions: (i) how long can the way she/he earns a living from the land be sustained; (ii) what can be done to make the method of land use sustainable if it turns out not to be, and (iii) will intervention to create sustainable forms of land use lead to a profitable enterprise (with 'profitable' defined in terms of the user's objectives)?

---

1. The term 'land user' refers to all parties who can be shown to have some legitimate claim to the land. The term is not only restricted to the direct user such as the pastoralist, but also considers other interests in society such as members of the local community, and those concerned with the national economy or with the global ecology. Each one of these interested parties has different, and often conflicting, management objectives. They are also concerned with different time (or sustainability) horizons. Attributing the right weights to each claim, and sorting out conflicts which may arise, is the subject of political and not physical science.

Agriculture in mountainous regions is a case in point. In such environments, soil conservation is usually geared to dealing with the *natural* causes of soil erosion. Humans have decided to control soil erosion not because it is caused by their method of land use, but because by controlling these natural processes of land degradation, humans are able to engage in a profitable economic activity which can be sustained for as long as it matters to them. In some cases, agriculture combined with soil conservation reduces rates of soil erosion to levels below those occurring under natural vegetation - for instance, in the case of terraced agriculture in Nepal and Java.

Thus, in considering whether a particular carrying capacity is sustainable, we must consider all processes which potentially lead to an effective decline in rangeland productivity, irrespective of the cause.

A methodology for evaluating land degradation

Principles

The discussion in the previous section conditions the way in which the sustainability of livestock production must be evaluated. Firstly, land degradation must be considered in terms of effective permanent production declines. Secondly, account must be taken of both the objectives of the land users and the socio-economic environment within which they operate and lastly, no distinction should be made between 'natural' and 'human-induced' processes of degradation.

Ideally, a rangeland monitoring strategy is designed which allows decision-makers to decide when to intervene. However, even well-designed long-term monitoring programmes are bound to fail for the following reasons:

(i) As time progresses, land uses and management objectives change. It is therefore quite likely that what is being monitored today will soon become irrelevant. This has been the case, for instance, with the ill-fated rangeland monitoring programme in Botswana.

(ii) It is almost impossible to distinguish irreversible long-term degradation from reversible short and medium-term production declines such as are caused by climatic fluctuations, and changes in land management.

The method proposed here relies on ecosystem analysis and modelling to simulate rangeland behaviour. Scenarios of future production (production forecasts) are used to form the basis for decision-making.

Forecasting

The methodology restricts itself to forecasting rangeland productivity as conditioned by processes of land degradation which act upon it, and proceeds as follows, for a study area:

(i) processes of environmental change are identified which potentially lead to an effectively permanent decline in rangeland productivity;

(ii) a model is then designed, calibrated and validated (if possible) to forecast future production as affected by those processes of land degradation;

(iii) the forecast is then interpreted in terms of intervention needed.

The special case of rangelands

Rangeland is potentially affected in two ways by the processes of degradation. Vegetation can be affected by the process of bush encroachment and by a decrease in the quantity and quality of the herbaceous cover, both of which are common occurrences on heavily stocked rangeland. The capacity of the soil to produce plant biomass is affected by a number of degradation processes: physical (e.g. surface capping and erosion), chemical (e.g. salinisation) and biological (e.g. the mineralisation of organic matter).

Current handbooks on rangeland management include numerous economically-viable ways of dealing with most changes in the vegetation. Vegetation changes are therefore tentatively considered as *reversible* processes of change, and are not here interpreted as land degradation.

A number of soil degradation processes, such as soil erosion, are effectively irreversible in rangelands. As the soil erodes, the capacity of the land to retain water and nutrients declines, which may lead to a decline in primary production, thus affecting livestock production. Given the very slow rates of soil formation, this process is effectively irreversible, and is interpreted as land degradation.

The 'soil life', 'residual suitability' and 'production half-life'

Assessing the sustainability of high stocking densities involves a study of how soil erosion affects future production from rangeland. The concepts of 'soil-life' or 'residual suitability' and 'production half-life' can be used to estimate the length of time (in years) a given ('minimum allowable' or 'half the present') output from the land can be guaranteed (Elwell and Stocking, 1984; Biot, 1988; Biot et al., 1989).

The derivation of these measures of sustainability for the case of semi-arid rangeland is outlined in the next section. The method is illustrated using a case study in Botswana which has been reported in more detail elsewhere (Biot, 1988; 1990b; 1991). The potential further applications of this modelling technique are also illustrated with some theoretical examples.

The residual suitability of rangeland

A simple experimental approach

Biot et al. (1989) have outlined a simple experimental approach to the estimation of future production as affected by land degradation processes. The field technique used to collect the necessary data is based on the principle of sequential testing. First of all, sites within the landscape are identified, which are representative of the different stages of the degradation process being considered. Then both the land's productive capacity, and the rate of degradation are measured for each site.

Figure 10.1: Graphical derivation of production forecast from simultaneous production and erosion measurements on rangeland

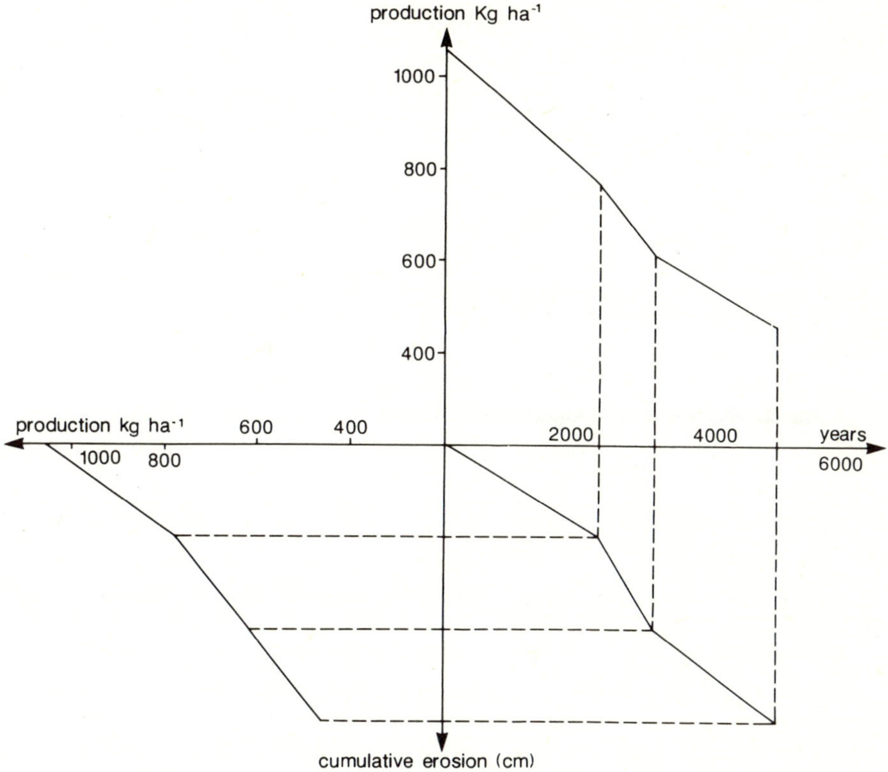

If such representative sites cannot be found, these can be simulated mechanically (see Stocking, 1985 for the case of soil erosion). A graphical technique is used to derive production forecasts from the data obtained (Fig. 10.1). The Soil Life is derived from this projection, based on a minimum allowable productivity determined by socio-economists.

A modelling approach

The estimation of the Soil Life can also be based on a fully mathematical model of the land's productive potential, such as EPIC (Williams et al., 1985) inclusive of inputs and outputs.

In a first attempt at developing simple models for use in regions which are not well endowed with banks of scientific information, Elwell and Stocking (1984) derived the soil's future production using:

(i) soil depth as an index of the soil's productive potential,

Figure 10.2: Derivation of the soil life according to Elwell and Stocking (1984)

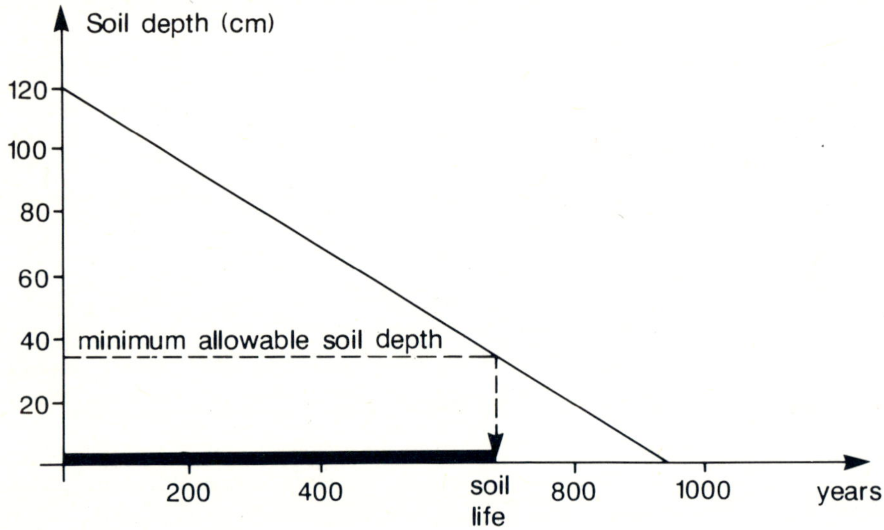

(ii) constant rates of erosion (i.e. not affected by changing levels of plant biomass and soil erodibility) and

(iii) constant rates of soil formation (i.e. equal to rock weathering alone).

The Soil Life is derived from the above projection. A minimum allowable soil depth is based on the land's ability to hold water, the frequency of significant rainfall, the relationship between plant biomass production and water availability (derived from work by Doorenbos and Kassam, 1979) work and finally a minimum allowable crop production level as given by socio-economists (Stocking and Pain, 1984), see Fig. 10.2.

Biot (1988) has assessed this estimation procedure in the light of a detailed analysis of the soil/landscape/plant system, and devised a modelling approach which uses:

(i) the Available Water Storage Capacity (AWSC) as a measure of the soil's productive potential. This is not only dependent upon soil depth (or the 'quantity' of soil), but also upon the 'quality' of the soil (soil texture, mineralogy and organic matter) as reflected in its Available Water Capacity (AWC);

(ii) rates of erosion conditioned by soil productivity: slow at high levels of productivity (i.e. dense ground cover), and much more rapid at low levels of productivity (i.e. sparse ground cover);

Figure 10.3: EPROM

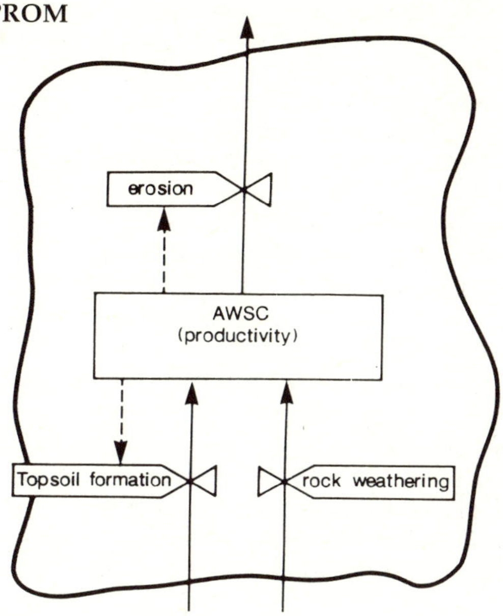

(iii) rates of soil accretion through rock weathering and organic matter production, the latter related to soil productivity.

These principles have been translated into a simple mathematical model, EPROM (Erosion PROductivity Model) which simulates future rangeland productivity (Fig. 10.3). THEPROM (Biot, 1990a) is a more complex model which enables the impact of different management scenarios to be studied.

The Botswana case study

Background

Botswana is similar to other regions in sub-Saharan Africa, in having experienced a steady increase in livestock numbers during the last 50 years. These numbers have now apparently reached a plateau, and fluctuate somewhat because of varying climatic conditions. The most common livestock production strategy is that of subsistence pastoralism, linked to a market economy; livestock sales in Botswana's case are through a well organised meat marketing board which sells beef at favourable rates to the European Community. Livestock management typically involves high stocking densities on predominantly communal rangeland.

The savanna vegetation has undergone profound changes, as livestock have replaced the original herbivore and carnivore populations (Campbell and Child, 1971). In the light of these changes, the sustainability of the present level of livestock production has been questioned by numerous scientists (see Arntzen and Veenendael, 1986, for a comprehensive review of environmental issues in the country). The Botswana government has often been advised to design schemes which would decrease the livestock pressure on the land. Abel and Blaikie (1989) discuss the appropriateness of these schemes.

As has unfortunately been the case in many conservation programmes, the need and urgency for government intervention has been mainly derived intuitively by people trained in widely differing ecological and economic environments, rather than derived from evidence collected under rigorous scientific procedures. The question as to exactly how sustainable high stocking densities are in Botswana has largely remained unanswered. This question is addressed here using the methodology outlined earlier. The study is restricted to the hardveld landscape in the South-Eastern part of the country where livestock is concentrated.

The forecast

A geomorphological and soil survey of a rangeland tract over granitic rocks typical of conditions in the hardveld revealed that sheet and rill erosion is the most widely occurring and potentially the most important process of land degradation (as defined earlier) in this region.

The effect of sheet and rill erosion on the productive capacity of the range was assessed through EPROM. It was hypothesised that the way in which soil erosion affects rangeland productivity in this dry environment is through its impact on the soil's capacity to store water (AWSC: AWC x soil depth).

A survey showed that standing biomass of the herbaceous layer is a function of the AWSC. The rate of soil erosion was estimated using the Soil Loss Estimation Model for Southern Africa (SLEMSA) (Elwell and Stocking, 1982), after calibrating soil erodibility and rainfall erosivity based on the results of a water balance study in Mahalapye, and on sediment traps scattered within the survey area. More details about the calibration procedure can be found in Biot (1988). Soil erosion affects the AWSC through a reduction in soil depth, but also by selectively removing from the soil those particles most conducive to a high AWC: namely, clay and organic matter. Enrichment ratios for both these soil particles were derived from natural sediment basins, and a relationship established empirically between clay and organic matter and AWC.

The survey was completed in two stages: a 6-week reconnaissance survey to identify the main processes operating in the system, followed by a 6-month survey during the main rainy season. Prior to their use in the calibration of EPROM's input and output functions, the data were adjusted for 'average' rainfall conditions using rainfall/vegetation and rainfall/erosion relationships derived for similar environments in Zimbabwe (Dye and Spear, 1982; Elwell and Stocking, 1982).

The simulation of future rangeland productivity in the test area is reproduced in Fig. 10.4. The forecast reveals that, under the prevailing grazing intensity (6.6 ha/LU), rangeland productivity declines, confirming earlier projections by rangeland scientists. After an initial slowing down of the rate of degradation (which is caused by an increase in tree cover), the pace of productivity decline accelerates. This is caused by a decline in herbaceous

Figure 10.4: Productivity forecast using EPROM

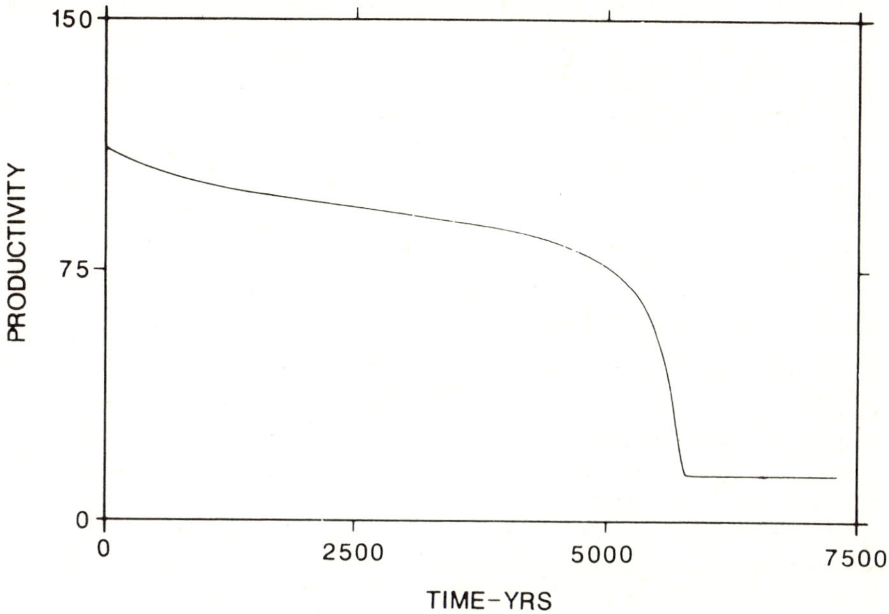

cover in response to the declining soil productivity. At a very low level of soil productivity (i.e. very shallow soil), the system stabilises as it is protected against further erosion by the emergence of gravel and rock fragments.

The residual suitability of rangeland in the test area

The relationships between soil productivity, standing grass biomass at the end of the season and stocking densities at the time of the survey have been used to estimate standing grass biomass at the end of the dry season for each level of soil productivity. This was corrected for the rainfall conditions which prevailed at the time of the field survey compared to average amounts of rainfall.

The following function relates standing biomass at the end of the dry season to soil productivity (AWSC: Available Water Storage Capacity, in mm):

$$\text{AWSC} = e^{2.591 \ln(y/39.9)}$$

with y = standing grass biomass at the end of the dry season.

The level of grass production needed to sustain the present day animal population has been calculated on the basis of a consumption rate of 2% of the weight of a tropical livestock unit (350 kg) per day, and a grazing season of 10-11 months. Using a utilisation coefficient of 0.70-0.75, and an average stocking density equal to 6.6 ha:LU, we calculated a minimum grass standing

Table 10.1: Soil Life of rangeland soils of different productivity levels on the eroding pediment slopes of a granitic landscape in the hardveld of Botswana

Soil productivity (mm AWSC)	*Soil life (years)*
120	6,113
90	2,363
60	394
30	131

Note: With an AWC of 10 % v/v, the figures in mm AWSC are equal to soil depth in cm.

biomass at the end of the dry season of 88-125 kg/ha. The minimum soil productivity level needed to produce this amount of grass has been called the 'minimum allowable soil productivity'. In our example, it ranges from 7.8 to 19.3 mm, or an average of 13.6 mm.

The time taken for soils of different productivity to reach this 'minimum allowable soil productivity' can be calculated using EPROM. The results are summarised in Table 10.1.

The average soil productivity level of the region has been derived from the soils map. The average Soil Life is equal to the time taken for a soil of average depth to reach the minimum allowable soil productivity. In this case, it amounts to 420 years. It is concluded that although land degradation processes occur in the test area, they are not likely to significantly influence pastoralists' ability to earn a living from the land within a time-frame which is relevant to them, and that the stocking densities imposed on the land are indeed sustainable.

Further applications of this approach

The impact of various interventions on present and future rangeland productivity can be assessed through THEPROM (Fig. 10.5). The model requires many more parameters to be estimated. Moreover, there are problems of how the evolution of the tree cover is related to erosion, which have not yet been adequately incorporated into the model. Application is therefore restricted at present to the study of how a pure grassland responds to management.

In the following example, the response to management of grassland under two different climatic regimes is simulated.

The two climates are characterised as 'semi-arid' and 'wet and dry tropics'. The characteristics of the semi-arid climate, and the derivation of the parameter values for THEPROM are given elsewhere (Biot, 1990a). The

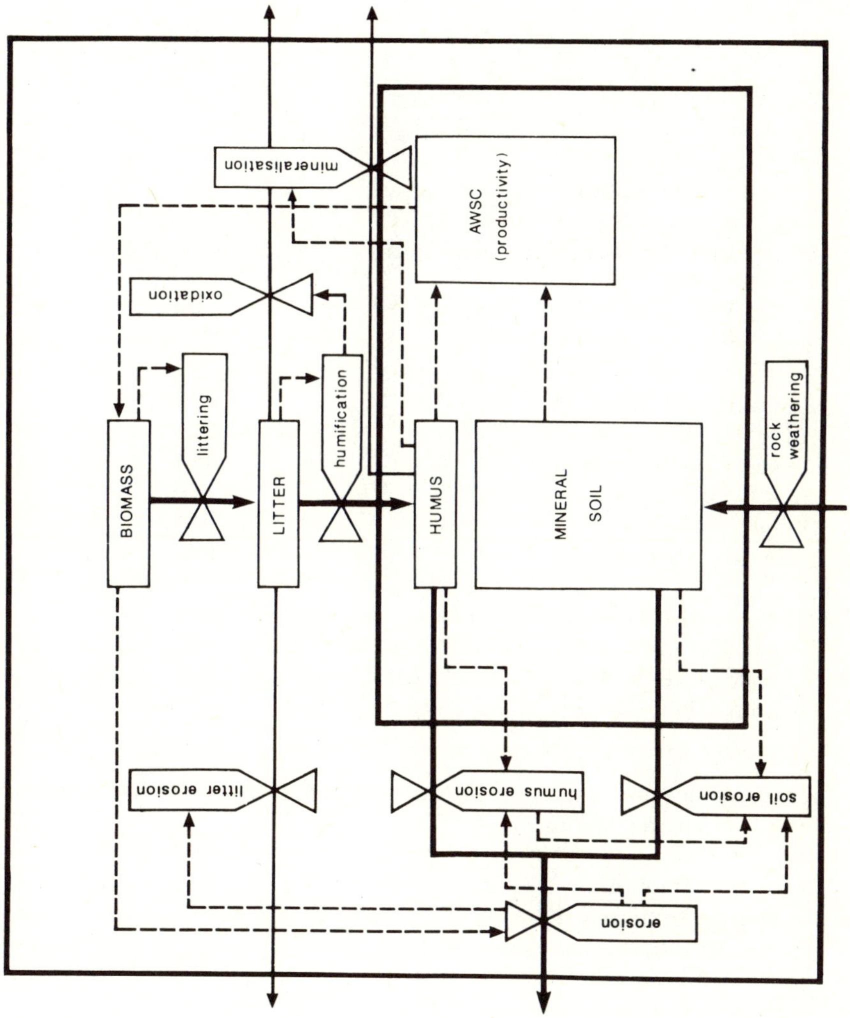

Figure 10.5: THEPROM

Table 10.2: Parameter values of THEPROM used to simulate the response of grassland under two different climatic conditions

Parameter	*Semi-arid*	*Wet and dry tropics*
Soil depth	120 cm	120 cm
Bulk density	1.5 g/cm^3	1.5 g/cm^3
AWC mineral soil	10 %	10 %
AWC/%om	5 %	5 %
Biomass production	2,500 kg	5,000 kg
Litter cover	2,000 kg	4,000 kg
Organic matter	0.5 % over 10 cm	1.0 % over 10 cm
Mineral soil formation	100 kg/year	400 kg/year
AWC/organic soil coef.	0.000333	0.000333
AWC/mineral soil coef.	0.00000667	0.00000667
Soil erodibility	3	4
Rainfall energy	8700 J/m^2	22560 J/m^2
Slope length	200 m	200 m
Slope angle	2.5 and 15 %	2.5 and 15 %
Minimum soil depth	15 cm	15 cm
Cover/biomass coef.	10.7	21.8
Cover/biomass constant	30.4	95.9
Litter erosion coef.	0.05	0.05
Mineralisation coef.	0.038	0.042
Oxidation coefficient	0.8	0.8
Humification coefficient	0.8375	0.8810
Litter production coef.	0.7146	0.7297
Enrichment ratio	4.5	3
Biomass prod./AWSC coef.	254, 0.0787	254, 0.0787
Maximum biomass prod.	2,550	5,100

grassland in the wet and dry tropics differs from the semi-arid in the following characteristics:

- higher rainfall,
- higher biomass production,
- higher litter production,
- higher organic matter content of the soil,
- higher rate of mineral soil formation,
- 'better' soil erodibility,
- higher rainfall erosivity,
- more rapidly responding crop cover/biomass function,
- more rapid mineralisation, humification and litter production,
- lower enrichment ratio.

The parameter values used are given in Table 10.2. In Fig. 10.6, future rangeland production is forecast for both climatic conditions without animals. Fig. 10.7 illustrates the influence of animals on the system under wet and dry climatic conditions. The main features of these simulations are summarised in Table 10.3. Using Table 10.3, we deduce that the presence of animals shortens the system's lifespan by about 15-20% in the wet and dry tropics, and by about 30% in the semi-arid areas.

Table 10.3: Summary of simulation results – wet and dry tropical and semi-arid rangeland

Climate	*Semi-arid*				*Wet and dry*			
slope (%)	2.5		15		2.5		15	
stocking rate (LU:ha)	0	1:5	0	1:5	0	1:5	0	1:5
initial gross soil loss (mm/year)	0.15	0.16	1.47	1.66	0.15	0.16	1.47	1.66
initial net soil loss (mm/year)	0.14	0.16	1.46	1.65	0.12	0.14	1.44	1.63
system's life (years)	4992	3616	468	340	4240	3408	342	282

Note: Gross soil loss: soil erosion
Net soil loss: soil erosion - soil formation

The simulation results also illustrate how slope affects the sustainability of the system: for the case of the wet and dry tropics, an increase in slope angle from 2.5 to 15 % causes a drop in the system's soil life which is much more spectacular than the introduction of high numbers of livestock.

The model allows the study of how the system responds to changing management intervention. Fig. 10.8 illustrates this for the steeply sloping range in the wet and dry tropics. In this figure THEPROM is first run for 150 years with an animal density of 1 LU:5ha. The animals are then removed, and the system run for another 50 years. When this does not seem to stop land degradation sufficiently, a concerted soil conservation effort is put in which involves shortening the slope length to 20m by introducing strips of permanent vegetation, and the improvement of soil infiltrability (soil erodibility = 4.5) by mechanical means; animals are allowed back on the land, but at lower densities (1 LU:10ha).

Figure 10.6: Productivity forecasts in the absence of livestock for semi-arid (a) and wet-and-dry tropical (b) conditions using THEPROM – litter, biomass and organic matter (OM) in kg/ha; soil in cm

(a)

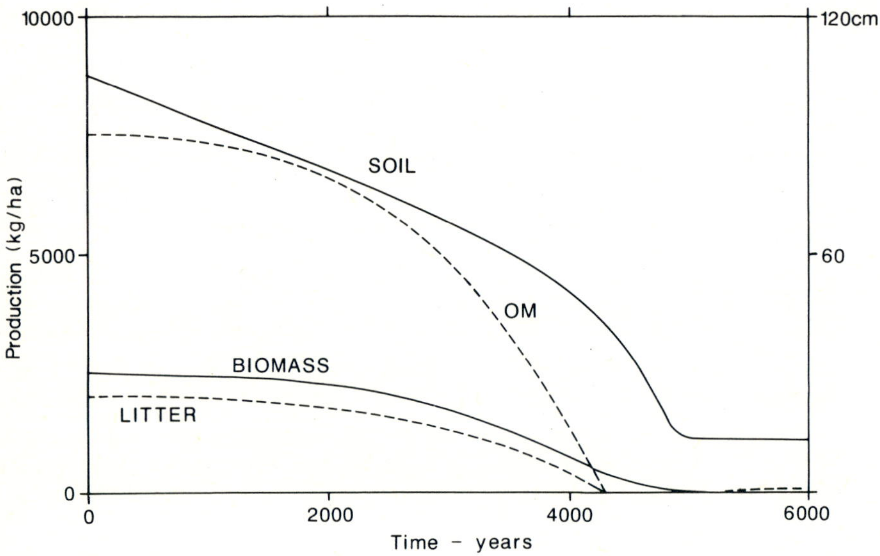

(b)

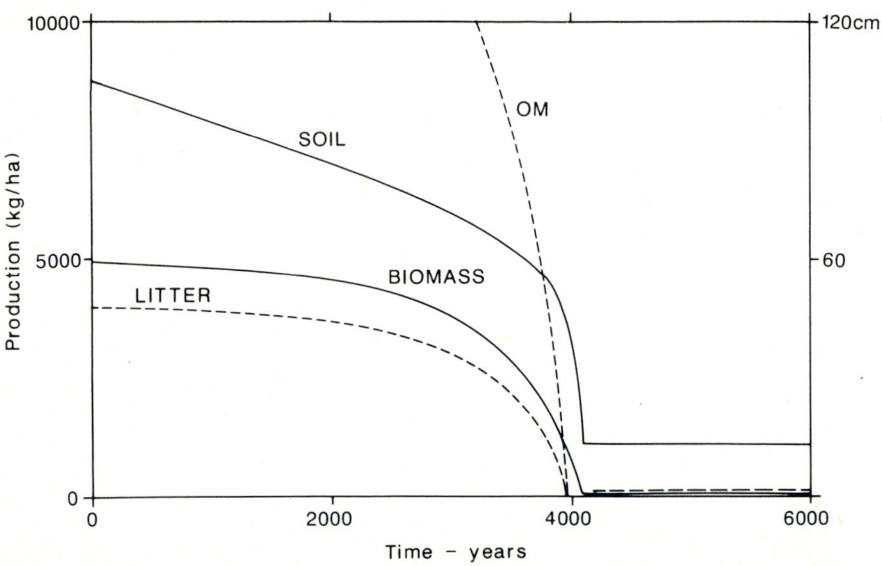

Figure 10.7: Productivity forecast in the presence of livestock for wet-and-dry tropical conditions using THEPROM

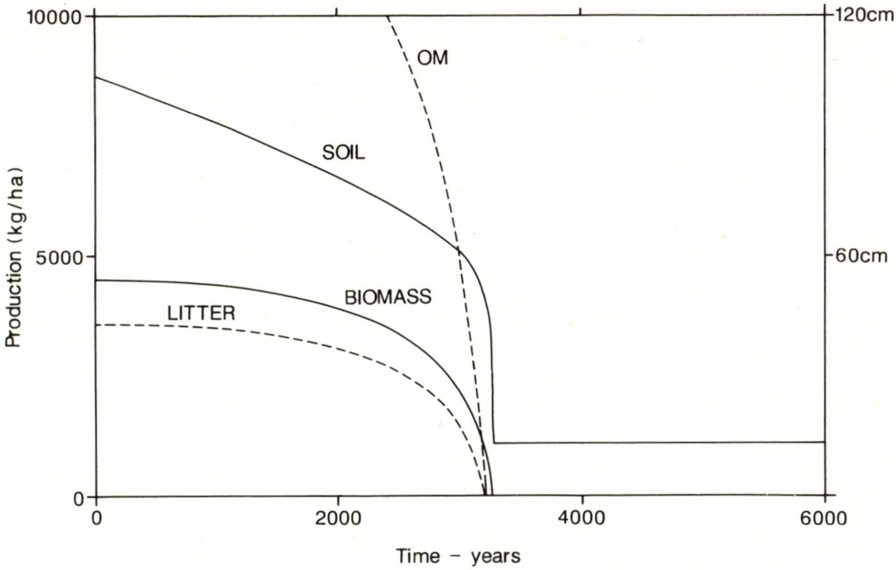

Figure 10.8: Response of rangeland productivity in the wet-and-dry tropics to intervention

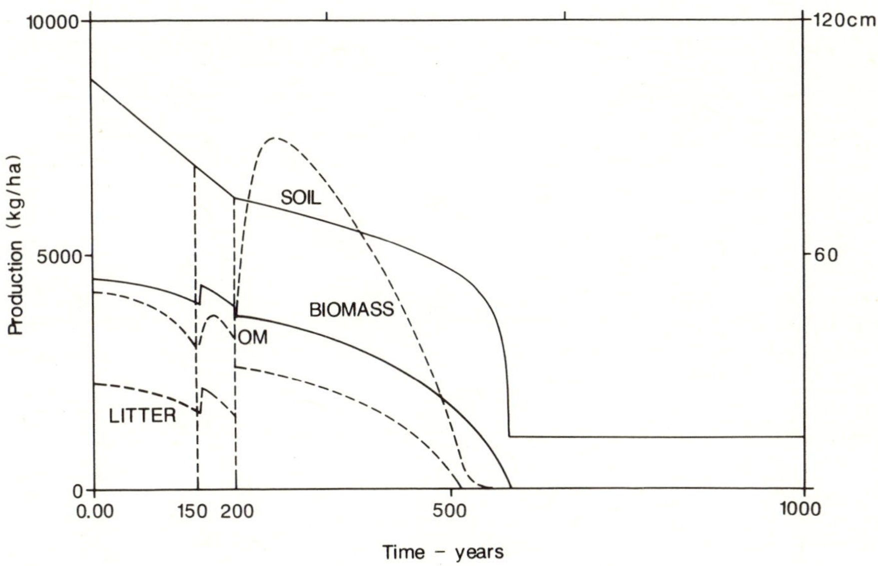

Discussion

Carrying capacity and soil life

The simulations of rangeland behaviour using THEPROM illustrate how the sustainability of livestock production is related to stocking density. This relationship can be used to estimate the maximum stocking density which can be sustained for a pre-determined 'sustainability horizon' - say, 100 years. Fig. 10.9 illustrates the procedure for a soil of 60 cm, using THEPROM (see also Biot et al., 1989).

Preserving the resilience of the vegetation

Thus far, THEPROM has been used to simulate the effect of soil degradation on rangeland productivity. It has been mentioned earlier that vegetation can also undergo change to the extent that rangeland productivity is affected. Up to now, these changes have been assumed to be reversible, and were not therefore regarded as rangeland degradation.

This view is not shared by all rangeland scientists. De Leeuw and Tothill, in this book, list a number of authors who contend that a minimum amount of biomass must be allowed to stand at the end of the dry season for the long-term productivity of the herbaceous cover not to be adversely affected. It would require less time for the rangeland to reach this minimum allowable grass cover, necessary for reasons of resilience and preservation of species

Figure 10.9: Relationship between stocking density and soil life and derivation of the carrying capacity for a given 'sustainability horizon'

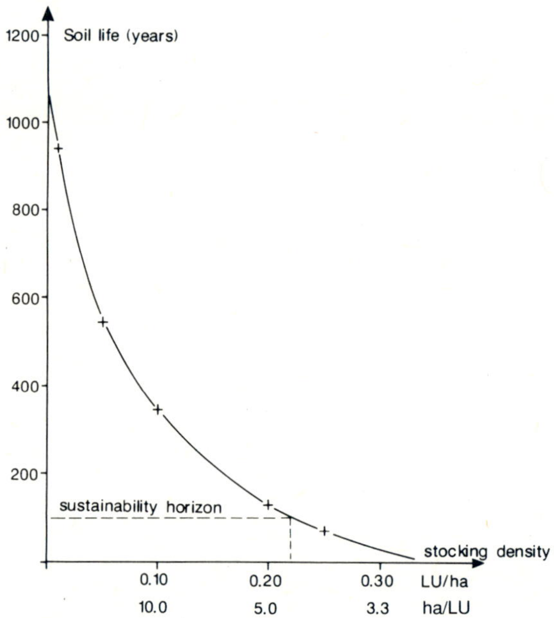

Table 10.4: Soil Life of rangeland soils of different productivity on the eroding pediment slopes of a granitic landscape in the hardveld of Botswana (corrected for reasons of preservation of the productive potential of the sward)

Soil productivity (mm AWSC)	*Soil life (years)*
120	5,644
90	1,875
60	exceeded
30	exceeded

diversity in the herbaceous cover, than if we let the rangeland become depleted entirely. This can be estimated using EPROM, for instance, as follows: based on equation (1) above, and using a minimum allowable amount of biomass at the end of the dry season equal to 200 kg/ha (Wylie et al., 1988), we derive a minimum allowable AWSC = 65 mm. The 'soil life' is now as follows (Table 10.4).

We can now provide a more cautious recommendation with regard to rangeland utilisation. Again, using THEPROM, the feasibility of different stocking densities can be simulated as in Fig. 10.9, and the 'carrying capacity' of the rangeland can be deduced for different soil depths and 'sustainability horizons'.

Problems of uncertainty

Biot (1991) discusses the different types of uncertainty which are associated with this type of assessment. They are found in model design, measurement and sampling errors, and mean that the simulations cannot be viewed as true predictions: they can only be used as *guidelines*, and need to be interpreted carefully.

One way of assessing the effect of uncertainties on model behaviour is by

Table 10.5: Sensitivity of EPROM to a 20 % deviation (either side) from the mean input values used in the Botswana case study (soil life in years)

Variable	*Lower*	*Mean*	*Upper*
Present-day productivity	265	420	768
Standing grass biomass	318	420	∞
Minimum allowable productivity	363	420	∞
Productivity inputs	385	420	485
Productivity outputs	400	420	440
Productivity level	380	420	423

carrying out a sensitivity analysis. The sensitivity of EPROM to input data uncertainties is illustrated in Table 10.5.

In this case it is concluded that the model is fairly robust to measurement and sampling errors. The soil life is certainly in the order of several centuries, which puts it beyond the time frame relevant to today's pastoralists.

Problems of scale

In their review of the concept of carrying capacity, de Leeuw and Tothill, in this book, also raise an important issue of scale. In the case given above, only one landscape unit has been considered: the eroding pediment slopes, which, though occupying a large proportion of the total landscape, are not representative of the total rangeland productivity. Table 10.6 illustrates how different parts of the landscape carry different types and amounts of biomass, all of which are accessible to livestock as long as rangeland is communally owned.

Table 10.6: Vegetation characteristics and stocking densities in four different landscape units of the rangeland considered

	Soil/Landscape unit			
	I	*II*	*III*	*IV*
Trees				
Density (/ha)	2230	1404	634	2165
Canopy cover (%)	43	17	9	71
Mean height (m)	1.4	1.1	0.8	2.2
Herbaceous layer				
Density (/ha)				
grass	30	88	58	74
forbs	25	21	102	23
seedlings	17	81	108	18
Grass basal cover (%)	1.8	4.4	5.2	1.5
Bare soil (%)	76	66	62	75
Standing litter (kg/ha)	2233	538	434	931
Grass standing biomass (kg/ha)	228	698	346	192
Forbs standing biomass (kg/ha)	128	98	127	126
Annual grasses (%)	22	14	6	59
Stocking density				
Dung (/100m)	5.6	3.6	11.8	0
Cattle tracks (/100m)	1.5	1.6	2.2	2.0

Notes: Landscape units

I. Upper and mid pediment slopes on granites,
II. Upper and mid pediment slopes on granites/dolerites
III. Wide and shallow, sandy depressions
IV. Levees.

Characteristics of the herbaceous layer: mean for 2 samples (December and March) of the 1986/87 season.

These landscape units not only differ in their ability to produce grass: they also erode differently. Moreover, when considering erosion, some parts of the landscape act as sinks for the sediment produced elsewhere. Both the levees (unit IV), and the shallow sandy depressions (unit III) in Table 10.6 are such 'sinks'. The latter unit is also characterised by high amounts of biomass, and attracts large numbers of grazing animals.

Abel and Stocking (1987) estimate the *net* rate of soil loss from the hardveld landscape of Botswana to be only 20-25% of the rate of erosion derived using SLEMSA. Thus, the conclusions arrived at by looking at the eroding landscape units in totality exaggerate the danger from soil erosion. To get a clearer picture, one has to assess the redistribution of soil (and soil productivity) within those parts of the landscape used by the animals.

It is not yet clear how this problem can be addressed within the confines of routine rangeland resource assessment exercises. As far as erosion goes, the concept of 'Sediment Delivery Ratio' proposed by Abel and Stocking (1987) is not sufficient. In addition, data is needed on the net accumulation and type of sediment, together with an estimation of the surface areas covered by sedimentation and sediment 'routing'. When considering extensive rangeland, it is also necessary to know the sizes, probability of occurrence and accessibility to an average herd of different landscape elements.

Conclusions

In their chapter in this book, on the concept of carrying capacity, de Leeuw and Tothill argue that, although the traditional derivations of the concept have limited utility for assessing extensive rangelands, the underlying principles on which the concept is based need to be taken into account for sustainable resource management.

One of the problems with most estimation procedures of carrying capacity is that, though the need for sustainable stocking densities is stressed, there is little said or written about how this can be assessed. Recent attempts to assess sustainable stocking densities are based on a 'minimum allowable crop-cover' (Van Wijngaarden, 1985) or 'proper use cover' (Wylie et al., 1988); these covers would minimise the impact of grazing on the herbaceous cover of the following year. There have been very few attempts to look at much longer time-horizons (with the exception of the work by Dunne et al., 1979). The impact of soil degradation on future rangeland productivity, though widely acknowledged, has never been incorporated directly into the estimation of carrying capacity.

When considering the sustainability of livestock production, it is argued that only those processes which provoke an *effective* decline in rangeland productivity should be looked at. The initial cause of productivity decline is, at this level of analysis, of little relevance.

The concept of 'soil life' provides a framework within which the long-term productivity of rangeland can be estimated. It is now possible to assess the impact of different stocking densities on the capacity of the land to continue supporting a 'minimum allowable' number of animals required for a given level of earning from the range, in the prevailing socio-economic conditions.

Modelling the behaviour of rangeland on computer is preferred over a monitoring approach which has little to offer in the case of long-term rangeland degradation. Computer models, once suitably calibrated, allow simulation of rangeland response to a wide range of management interventions.

The case study has illustrated the potential application of this modelling technique. It has been shown, for Botswana, that the savanna is able to support present-day livestock densities for another 400 years. The maximum livestock density for a 100-year sustainability horizon has been set at about 5 ha/LU.

A number of uncertainties affect the outcome of the technique. These have been reviewed briefly in this chapter. A much more important issue is that of scale, which needs to be considered as a priority for future research.

11
Reducing Cattle Numbers on Southern African Communal Range: Is It Worth It?*

N.O.J. Abel

Introduction

This chapter is about the livestock 'carrying capacity' debate among developers, range scientists and pastoralists of the communal rangelands of sub-Saharan Africa. It concentrates on Southern Africa, with particular reference to Botswana, but some generalisations are made. It is a preliminary and very brief report of a model which relates stocking densities of cattle to their productivity and to the rate of soil lost from their range. The model is used to assess the change in output of livestock products, as well as changes in income and soil loss, resulting from a reduction in stocking density. It addresses the question of whether it worthwhile to reduce the numbers of livestock on communal rangelands in Southern Africa.

Pastoralists of sub-Saharan Africa mainly perceive their worst problems as being drought and insufficient livestock. Most of them try to promote rapid increase of their livestock, between droughts, and few attempt to limit numbers voluntarily (Sandford, 1983a).

Policy-makers, and the majority of range scientists, see the main problems of livestock development in sub-Saharan Africa as over-accumulation of stock, insufficient production of milk and meat for urban consumption and export, and land degradation. They assume that by lowering livestock density, the productivity of individual animals will increase, which will more than compensate for the reduced number of animals, and furthermore, that the reduction in density will conserve rangeland. Development policies built around this view require pastoralists to reduce animal numbers, and frequently also promote changes in land tenure towards group or private control (Behnke, 1985b; Cousins, 1987; Sandford, 1983a; Abel et al., 1987).

---

* The aerial surveys on which some of this work is based were sponsored by the Government of Botswana and the International Livestock Centre for Africa (ILCA), and I am grateful for their support. Thanks are due to Yvan Biot, Michael Stocking and Piers Blaikie of the University of East Anglia, who have given me a lot of time during the development of this work. Thanks are also due to the Applied Ecology Research Group, Arthur Georges, David Williams and Mark Carson in particular, and my other colleagues at the University of Canberra. I am very grateful to Judy Caughley for her constructive analysis of the approach, and to Mark Stafford Smith for comments on the manuscript. While acknowledging the helpful influences of these institutions and people, I am responsible for the views I have expressed.

The first section discusses the origin of the questions analysed by the model. The next two sections pass without detail over the technical aspects of the model, which are described in full in Abel, 1993. A few of the main results are then discussed, while the final section is a brief comment on some policy implications.

Destocking and the degradation issue

Pastoralists have not co-operated in de-stocking, neither in colonial nor in post-independence programmes. One reason is security against drought: on communal rangeland, where animals are privately held, in the absence of agreements among herdowners to de-stock, it is rational for herdowners to enter a drought with as many animals as possible, because the chance of a nucleus herd surviving is increased. Another reason is the fear among those who already lack sufficient animals that a reduction in animal density may mean that fewer people can be supported. A third reason is that pastoralists frequently do not accept that their pastures are degrading. A fourth may be the typically inequitable distribution of livestock among pastoralists: for those who already have insufficient, to reduce their livestock further would sink them into extreme poverty (Abel and Blaikie, 1989), while those who hold many animals are often politically influential and able to block de-stocking in their own cases.

There is little argument against the first of these reasons, which is not dealt with further here. The second and third reasons are discussed below. Discussion of the fourth, although crucial to the debate, is beyond the scope of this chapter.

The fear among pastoralists that de-stocking would reduce their food supply is supported experimentally. The relationships between stocking density and mass gain per livestock unit and per hectare are well established and should have been known to range scientists (Payne, 1990; Stoddart et al.,1975; Jones and Sandland, 1974; Hart, 1978). These relationships have, however, been largely ignored in the debate on carrying capacity (see for example Rennie et al., 1977; Prescott, 1961). They have been reintroduced to the debate by Behnke (1985a), and de Ridder and Wagenaar (1986). It is clear that, under experimental conditions, maximum mass gain per hectare is achieved at a higher stocking density than is the maximum rate per livestock unit (APRU, 1980; Barnes, 1965; Carew, 1976). Making the reasonable assumption that mass gain is a proxy for secondary productivity in general, it is apparent that reduction in density would decrease the number of people supported, despite the expected increase in the productivity of individual animals. The extent of this decline would depend on the specific characteristics of the animals and the range.

The third reason why pastoralists have not de-stocked was that they do not generally accept that their rangelands are degrading. One of the problems here is that opponents in the debate on carrying capacity usually hold different views on the nature of degradation. Thus a decline in the abundance of palatable perennial grass species may be seen as degradation or as change, according to one's viewpoint. Abel and Blaikie (1989:113) have defined range degradation as:

> ...an *effectively permanent* decline in the rate at which land yields livestock products under a given system of management. 'Effectively' means that natural processes will not rehabilitate the land within a timescale relevant to humans, and that capital or labour invested in rehabilitation are not justified...This definition excludes *reversible* vegetation changes even if these lead to temporary declines in secondary productivity. It includes effectively irreversible changes in both soils and vegetation [emphasis in the original].

Abel and Blaikie (1989) then stressed the great instability of most semi-arid vegetation, arguing that frequent and large changes are usual. The generally resilient nature of this vegetation means that many, but certainly not all, of the changes will be reversible. Warren and Agnew (1988), Scoones (1990) and Ellis and Swift (1988) are among a growing group of writers who reject the generalisation that communal rangelands are degrading through over-grazing. Meanwhile Le Houérou (1989) and Westoby et al. (1989) warn that it is dangerous to adopt the opposite assumption - that there is no such thing as effectively irreversible vegetation change. Clearly there is, but the general resilience of rangeland vegetation in sub-Saharan Africa means that soil losses or other detrimental changes in the soil are more likely than *direct* degradation of vegetation by grazing and browsing to be a widespread cause of range degradation.

Soil loss would cause range degradation if the rate of erosion were faster than the rate of soil formation. *Descriptions* of soil loss from rangeland are common (Blackburn 1984; Whitlow 1987), but there are few studies of the *effects* of erosion on primary production (Stocking and Peake, 1985). I have found no studies on the effects of erosion on secondary production. Kelly (1973) and Kelly and Walker (1976) have shown for Zimbabwe that, in a season of low rainfall, herbaceous production on range under heavy stocking was less responsive to rainfall than it was on more lightly stocked range. In a wetter year there was no difference in the response. The result was presumably due to a change in the soil that impaired moisture availability. A similar result was obtained by McKay (1968) from Botswana. Abel and Blaikie (1989) used Kelly's data to infer a negative relationship between soil depth and primary productivity. Biot (1988, 1990a and chapter in this book) had previously investigated the rates of soil loss, and the effects of this on rangeland degradation, in his study of communal pasture in Botswana. Soil depth was the major determinant of primary productivity on the granite landscape of his study area, but he concluded that, under the current stocking density, the rate of degradation due to soil loss was very slow - of the order of centuries before an appreciable decline in herbaceous primary production could be expected.

The need for modelling animal productivity and soil loss

Three important aspects of the carrying capacity debate have been raised by this discussion so far: the productivity of animals, the secondary productivity of land, and the rate of land degradation at various stocking densities. No study has attempted to integrate all three aspects. Moreover, existing studies on Botswana have a number of limitations, as will now be discussed.

Firstly, although there is clear indication that the productivity of livestock per hectare may decline with de-stocking, de Ridder and Wagenaar's (1984, 1986) analysis was a static one, based on a comparison between two unvarying stocking densities, one representing ranches in Botswana, the other communal areas. Sandford (1982b) has argued that the opportunistic strategies of pastoralists, in which numbers of animals fluctuate widely over time, give a total output which is higher in the long-run than the more conservative stocking strategies of ranchers. It is therefore necessary to investigate outputs from systems at different stocking densities over a span of time during which environmental conditions fluctuate widely.

Secondly, de Ridder and Wagenaar's model was very simple, using production coefficients from the literature, or assumptions backed by sensitivity analysis where no data were available. The model made no connection between production and environmental conditions. Thus, although their model has been extremely useful, it needs to be developed further by including these aspects.

Thirdly, the productivity of ranching was compared by de Ridder and Wagenaar (ibid.) with that of a communal system. This is very useful, in that ranching is held up as an ideal for livestock development in much of Africa, but assessing ranch productivity, under lower stocking densities, does not help us predict the effects of de-stocking on the existing communal system, with its higher stocking density. The effects of destocking are likely to differ, as management, access to the factors of production, goals and outputs of ranching and communal area production differ widely.

A further limitation is in Biot's model, where although the model of soil loss and land degradation was highly developed, the links between rainfall, grass production, grazing intensity and grass cover were not strongly established. My work on this aspect has shown that the evolution of grass cover over the course of a season has an important effect on the rate of soil loss, an aspect neglected by Abel and Stocking (1987), and Abel and Blaikie (1989), as well as by Biot (1988).

A final limitation is that none of the models has analysed spatial variation in the incidence of erosion, as related to topography, soil type and livestock distribution.

One other model potentially available is that of Konandreas and Anderson (1982). This was rejected on the grounds that its information requirements, using data on individual animals, monthly time steps and pasture quality, are immense. Moreover, land degradation is excluded. A further problem with this, as with any herd projection model, is that it is difficult to control the trajectory of the model, leading to unrealistic projections of animal numbers.

Very small adjustments to calving and mortality rates can alter final population densities by orders of magnitude.

Drawing on the above critique of existing models, the model introduced in the present chapter was developed according to the following design criteria:

(i) it should have modest data requirements;

(ii) there should be good links between ecological conditions and livestock production, and between experimental or survey results and estimated outputs; or explicit and testable assumptions should be made about these relationships;

(iii) densities of animals should reflect levels and variations of those levels seen in reality;

(iv) comparison of different stocking densities should be between production systems with similar methods, access to resources and production strategies;

(v) soil loss should be analysed in relation to the spatial distribution of livestock (that element of the model is not reported here);

(vi) production, stocking density and soil loss should be linked, so that the effects of de-stocking on output and on soil loss can be compared; this is one important way of deciding if de-stocking is worthwhile.

The construction of the cattle production sub-model is described in the next section, followed by a description of the soil loss sub-model. Results and discussion follow.

The livestock production sub-model

The subject of modelling is the 'traditional' mixed farming system of Eastern Botswana, in which cattle are important, as well as small stock and donkeys. Sorghum production is significant when the highly variable rainfall permits. Remittances have become more important than agriculture as a source of income for rural households, and much of this is invested in cattle (Kerven, 1982; Hubbard, 1986; Flint, 1986). The production sub-model was located in the Central Region (Fig. 11.1). In modelling this system, the arable component was excluded, except for the contribution of draught oxen to energy output from the herd and to the value of the crop. Also excluded were small stock, because insufficient information was available on their productivity. This sub-model therefore focuses on the cattle component of a more complex system.

The structure of the production sub-model is shown in Fig. 11.2. It estimates the productivity of cattle in the Region at two stocking densities: the existing density, called here the Current Stocking Rate (CSR), and an officially recommended stocking rate, or RSR. The characteristics of herds at the two densities are given in Table 11.1. The sub-model was run over the period 1978-88.

Figure 11.1: Locations of study areas

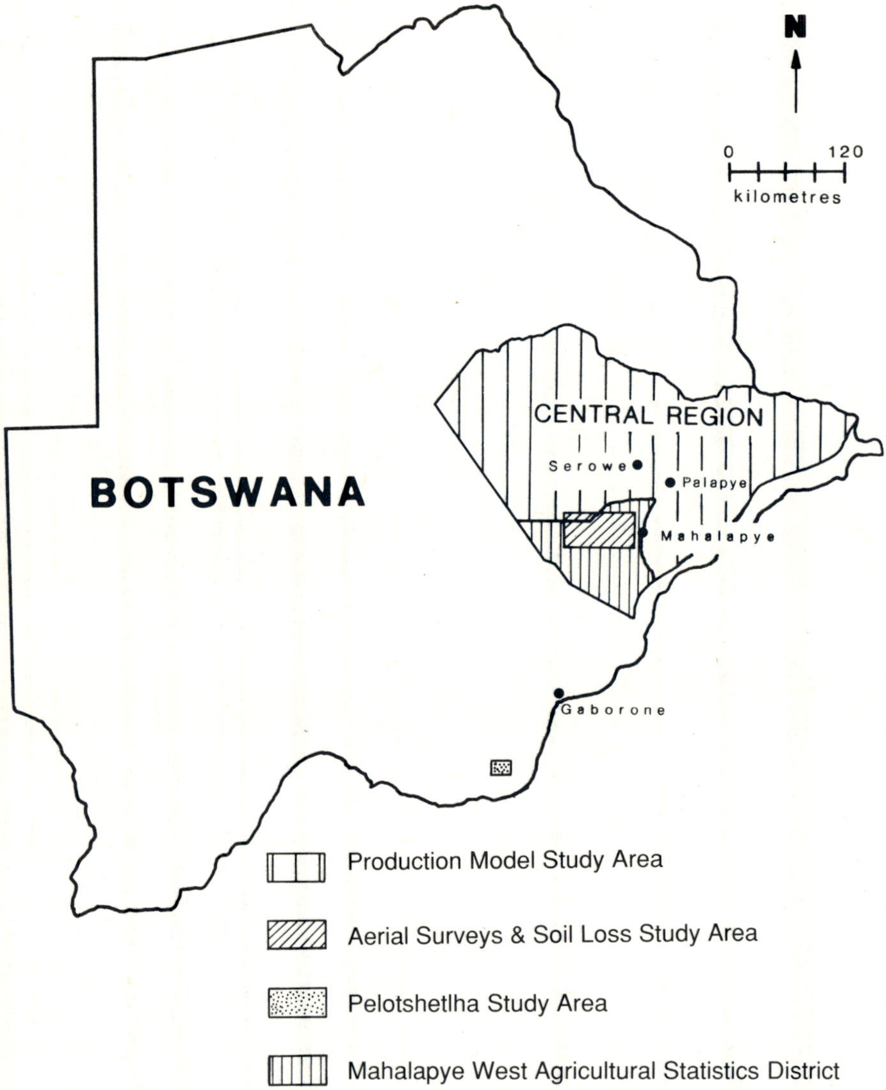

Table 11.1: Characteristics of herds at current and recommended stocking rates

	Current Stocking Rate	*Recommended Stocking Rate*
Calving rate, %	55.0	65.5
Calf survival, %	83.6	89.5
Proportion calves, %	19.8	25.4
Proportion cows, %	43.3	43.3
Proportion oxen, %	10.0	10.1
Stocking rate, ha/LU	6.4	14.2
12 m calf mass, kg	154.5	199.5
Available milk, kg/cow/d	0.9	1.3
Energy output, MJ/LU/y	667.8	876.9
Energy output, MJ/ha/y	105.0	62.1
Gross margin, Pula/LU/y	61.5	99.6
Gross margin, Pula/ha/y	9.7	7.0

Note: Pula is the currency of Botswana

The main driving variable of the production sub-model is rainfall. This determines variations in stocking density at the RSR, and calf growth and milk output at both densities. Other driving variables are the number, herd structure, calving and mortality rates of cattle. Driving variables are linked to production through functions estimated from published data, mainly the annual reports of the Animal Production and Range Research Unit (APRU, 1975-88). A curtailed description of the derivation of these functions follows. The equations and the measures of their precision are in Abel (1993). The relationships between rainfall and production are mainly described by linear or polynomial multiple regression equations. Slope coefficients and F values for the regression models are all statistically significant ($p < 0.05$). The level of precision varies between low and modest (the proportion of total variance in the regression models accounted for by rainfall is between 30% and 70%). This determined the need for the sensitivity analyses described in Abel (1993).

The following account of the derivation of equations linking production to rainfall begins with stocking densities, after which calving rate and calf survival, calf mass gain, milk production and off-take and energy for draught power are examined in turn. The description of the production sub-model ends with a discussion of the financial analysis of output.

Stocking densities

Central Region was selected for the production model because aerial surveys and land degradation studies had already been carried out there. The whole Region was selected, rather than a smaller unit within it, because, based on the population dynamics of livestock, the Region was judged to be an ecological unit, while smaller units were not.

Figure 11.2: Production model, Central Region, Botswana

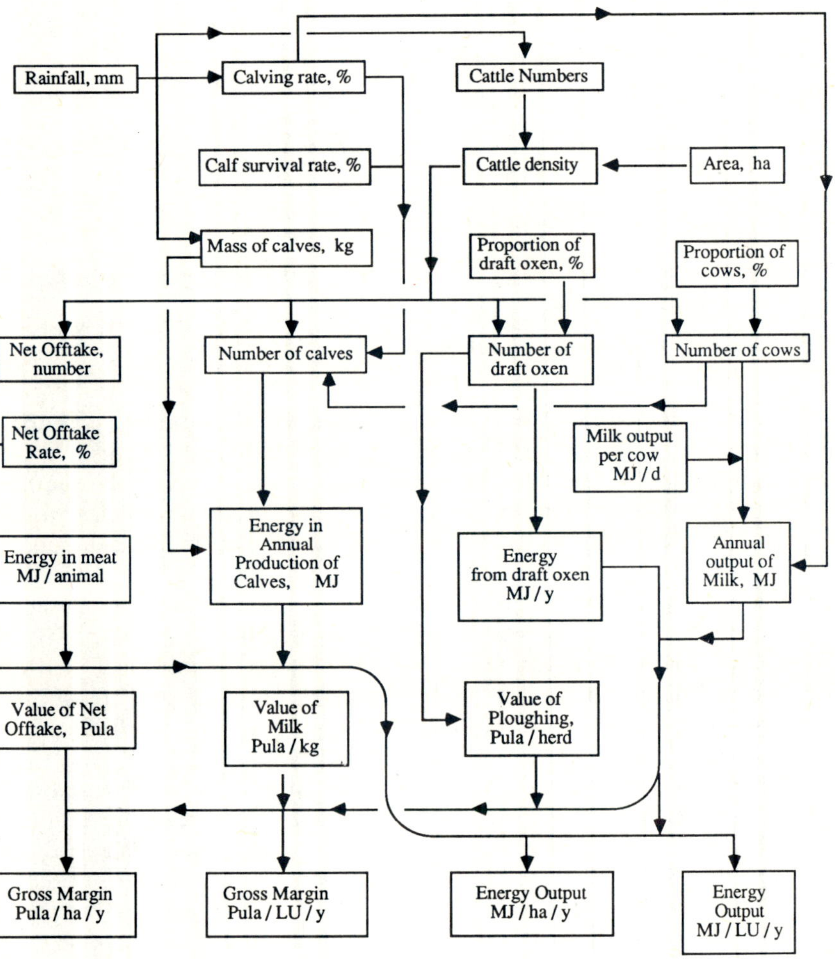

Not all of Central Region is grazed because not all is within reach of water. The extent of the area grazed was estimated as follows:

(i) the density of livestock in the aerial survey area in 1983 was estimated from aerial counts; it was assumed livestock density was the same over the rest of Central Region which was not flown;

(ii) Botswana Agricultural Statistics (Planning and Statistics Division, 1983) provided an estimate of livestock numbers in the region for 1983; this number divided by the density estimate (in LU/ha) gave an estimate of the extent of the area grazed;

(iii) I assumed the extent of the area grazed remained unchanged for the entire period of the study.

Densities in years when no aerial surveys were flown were calculated by dividing the number of animals, as recorded in Botswana Agricultural Statistics, by the number of hectares grazed.

Numbers were converted to livestock units (LU) using conventions established for Botswana by Carl Bro International (1982) in which one LU is 350 kg. Smallstock are 0.1 of a LU, donkeys 0.4. Flint (1986) gave weights and a herd structure which produced a mean mass of 248 kg for cattle at Pelotshetlha in South-east Botswana. Multiplying the number of cattle by 250/350 gives an estimate of cattle abundance in LU.

As already noted, although smallstock and donkeys are important in the present system (Flint, 1986), there is a lack of production data. Biomass of smallstock and donkeys is about 8% of the total biomass of domestic stock in Central Region. The biomass of smallstock and donkeys was converted to cattle biomass for the production model, and output adjusted accordingly.

Measures of abundance under the present system of management were provided by agricultural statistics (Planning and Statistics Division 1978-88). To obtain an officially approved stocking strategy (how stock densities should be adjusted to rainfall and the economic environment), published stocking densities from the Animal Production Research Unit research ranches around Botswana (APRU 1975-88) were regressed against rainfall for current and previous seasons. This equation was applied to the rainfall regime in the study area to produce variations in density over time. This variation was applied to the official carrying capacity figure for the region, which was derived from Field (1977), who used a parametric approach to set recommended stocking densities for 'Carrying Capacity' zones. Seven zones fell within Central Region. The density for each was weighted by the area of the zone and corrected for differences in Livestock Unit conventions used, to give a mean Recommended Stocking Rate (RSR) for the Region.

The mean densities for the eleven-year run are shown in Table 11.1, and fluctuations in Fig. 11.3.

Calving rate and calf survival

One of the indices favoured as a measure of ranch productivity is the energy contained in the output of calves (Rennie et al., 1977). Calving rates and calf survival under the Current Stocking Rate (CSR) were taken from the

Figure 11.3: Cattle stocking rates, Central Region, Botswana

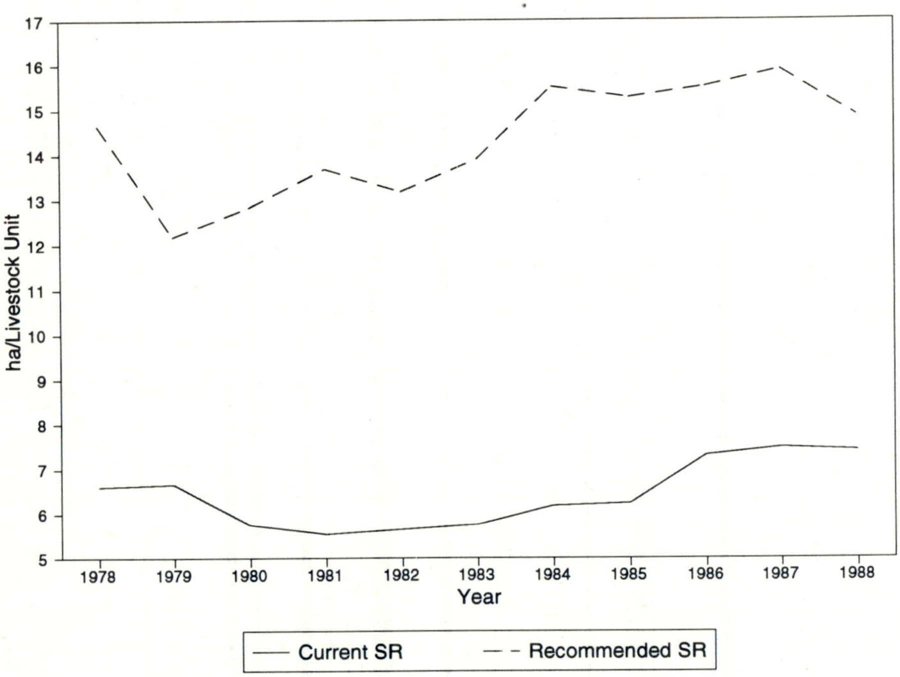

Agricultural Statistics (Division of Planning and Statistics). The level of calving rate and its variations for the RSR were derived by regression of calving rate against current and past rainfall, using data from research ranches (APRU, 1975-88).

There was no statistically significant relationship between rainfall and calf survival on the research ranches, so the mean ranch rate was applied in all years for the RSR system.

Mean calving rates for the eleven years are shown in Table 11.1. Sensitivity analysis suggested that RSR mean calving rate should be reduced to the level shown.

Calf weight gain

Ranch calves receive all their dams' milk and grow fast. In the 'traditional' system people take a proportion of the milk for themselves, and calf weight gain is retarded. Research data (APRU, 1975-88) provided relationships between past and present rainfall and weight gain for calves reared on ranches. This growth equation was modified to account for milk off-take and stocking density in the traditional system.

The effects on ranch calves of taking milk for humans has not been measured. Roy (1980) provides data from which an equation (the 'Roy equation') can be calculated which permits estimation of this effect. The equation estimates the amount of gross energy per kg of daily gain required for a given body mass of calf. Mass at twelve months without competition from humans for milk was estimated from the research ranch growth equation. The integral of the Roy equation is the gross energy requirement to grow a calf of that twelve-month mass. If some milk is removed, for humans, the amount by which calf mass at twelve months is reduced was calculated by transposing the Roy equation. In this way calf mass estimates for the RSR were produced, which varied with current and past rainfalls, and which were less than the mass gains on the APRU ranches thus allowing for human consumption of milk under the traditional system.

The next problem was to estimate calf gain in relation to rainfall under the CSR. It was noted above that Kelly and Walker (1976) from Zimbabwe, and McKay (1968) working on Morale Ranch within the Central Region, both offered evidence that grass production under heavy stocking may be similar to production under moderate stocking when rainfall is plentiful. When rain is sparse, grass production is very much less under heavy stocking than under moderate stocking for the same rainfall. This is taken into account in the production (and in the erosion) sub-model by assuming a steeper response of animal production to rainfall at the denser stocking rate. The rainfall-calf growth equation for the CSR was derived by calibrating it to the data of Rennie et al. (1977), who weighed calves from the traditional system between 1972 and 1974, and by using rainfall data from APRU publications. The rainfall-growth curve was sloped to reflect the steeper response to rainfall expected from the more densely stocked system.

Calf mass was converted to energy in mega Joules (MJ) using 50% as a killing out percentage, and meat energy content derived from Dahl and Hjort (1978).

Mean calf masses for the eleven years of the model run are given in Table 11.1.

Milk production and offtake

APRU (1986, 1987) recorded rainfall and measured milk production from communal area cattle which were fed supplements. The results supplied a relationship between rainfall and milk yield. This equation was used to estimate milk production at the RSR.

Flint (1986) measured milk production from communal area cattle at Pelotshetlha, in South-east Botswana. These cattle produced about 13% less milk than cattle in the APRU result (above) were estimated to produce at the same rainfall. Since cattle at Pelotshetlha are better managed than is generally the case in the communal areas, a milk production curve was set below the Pelotshetlha level, and sloped to converge with the RSR curve at 900 mm of rainfall. Milk for human consumption from lactating cows was then calculated as a mean daily output, over a year, of 0.9 kg/cow/day from the CSR system, and 1.3 kg/day from the RSR, over the eleven-year run.

Milk energy was derived from milk weight using a value from Konandreas and Anderson (1982).

The amount of milk produced is also influenced by calving rate. This influence was derived initially for the RSR from the relationship between calving rate and current rainfall derived from the research ranches. After sensitivity analysis, the RSR rate was reduced from a mean of 77% to a more likely 65%. Agricultural Statistics gave recorded rates for the CSR.

Milk production is further determined by numbers of cows in herds; this information was obtained from Agricultural Statistics, for the CSR. The proportion of cows in the RSR herds was assumed to be the same as for the CSR.

These characteristics of the herds are shown in Table 11.1.

Energy from draught oxen

The number of draught animals at the CSR came from Agricultural Statistics. Since the proportion of oxen in the herds was the same in the CSR and the RSR, the number of oxen at the RSR was derived from this proportion and the total number of cattle in that system. Energy output for ploughing was calculated using a figure from de Ridder and Wagenaar (1986).

Financial analysis

Values of animals and costs of production were obtained from the Planning and Statistics Division's Farm Management Surveys (1982-84), with milk and draught values obtained from Flint (1986). Net off-take of mature animals was used instead of calf mass gain in the financial analysis. The RSR net off-take rate was based on national commercial rates from Agricultural Statistics, regional data being unavailable. Agricultural Statistics provided net off-take rates for the CSR in Central Region. Prices and costs were corrected to 1982 levels. Results are expressed as gross margins - revenue less variable costs.

Sensitivity analysis

Sensitivity analyses were conducted on the effects on productivity of varying stocking densities, calving rate, milk yield, net off-take rate, calf mass gain, mean mass of animals in net off-take, and meat price. Only the calving rate under RSR emerged as having an unlikely value and a strong influence on productivity. A calving rate of 77% for the eleven-year run was derived, unmodified, from the relationship between rainfall and rate recorded on research ranches. This is probably much too high for communal area conditions in which *Brucellosis* is common. The rate was reduced to a much more plausible 65%. This was used in the runs which produced Figures 11.9 to 11.11. Table 11.1 gives the parameters used for these runs.

The erosion model

The structure of the erosion sub-model is shown in Fig. 11.4. The aim is to estimate rates of soil loss as an index of the rate of land degradation. This is an appropriate index for the area (Biot, 1988). The erosion sub-model used is SLEMSA - the Soil Loss Estimation Model for Southern Africa (Elwell and Stocking, 1982). This was calibrated in the study area by Biot (1988). SLEMSA requires data on slope percentage, slope length, an erodibility factor, effective cover, and rainfall erosivity. These are discussed after an outline of the general approach, which follows.

The study area for this model (Fig. 11.1) was selected because aerial counts of livestock and very low-level photographic surveys of the area had been made in 1982 and 1983. This data provided estimates of primary production and the number and distribution of livestock. Grass production data were needed for the development of a model relating rainfall to grass biomass, since effective vegetation cover is a major variable controlling soil loss. Estimates of animal numbers and distribution were needed so that the biomass consumed, and thus the cover removed, could be calculated, and so that the density of animals in the survey period could be calibrated to the Agricultural Statistics for the same period.

Livestock distribution was linked to spatial variation in rates of soil loss in the sub-model. Reporting this aspect is not feasible within the scope of the present chapter.

Topography

Aerial photographs (35mm colour diapositives), were taken at 2 km intervals along systematically aligned transects flown 5 km apart at 120 m above ground level. Only those plots falling unequivocally within a soil type were used - 144 plots in all. Slope at each plot was taken from 1:50,000 topographic map sheets. Slope length was assumed to be 30 m, a standard erosion plot. This yields an index of erosion, rather than a net figure which would be scale-dependent.

Erodibility factor

SLEMSA is very sensitive to the so-called F values - erodibility factors - which ought therefore to be calibrated. Biot (1988) did so for a soil on granite in the area. Stocking (pers. comm.) suggested values for the other soils and these were subjected to sensitivity analyses.

Effective cover

SLEMSA needs an estimate of the time-profile of vegetation cover over the rainy season, as the cover evolves from the end of the dry season in September to full cover by May. No relationship between rainfall and grass production was available for the study area. A grass production estimate was made by using the low-level aerial photos and livestock surveys. Marked experimental plots were photographed from the air at 120 m. The diapositives were projected on a screen and a subjective estimate of ground cover made. Cover

Figure 11.4: Soil loss and stocking density

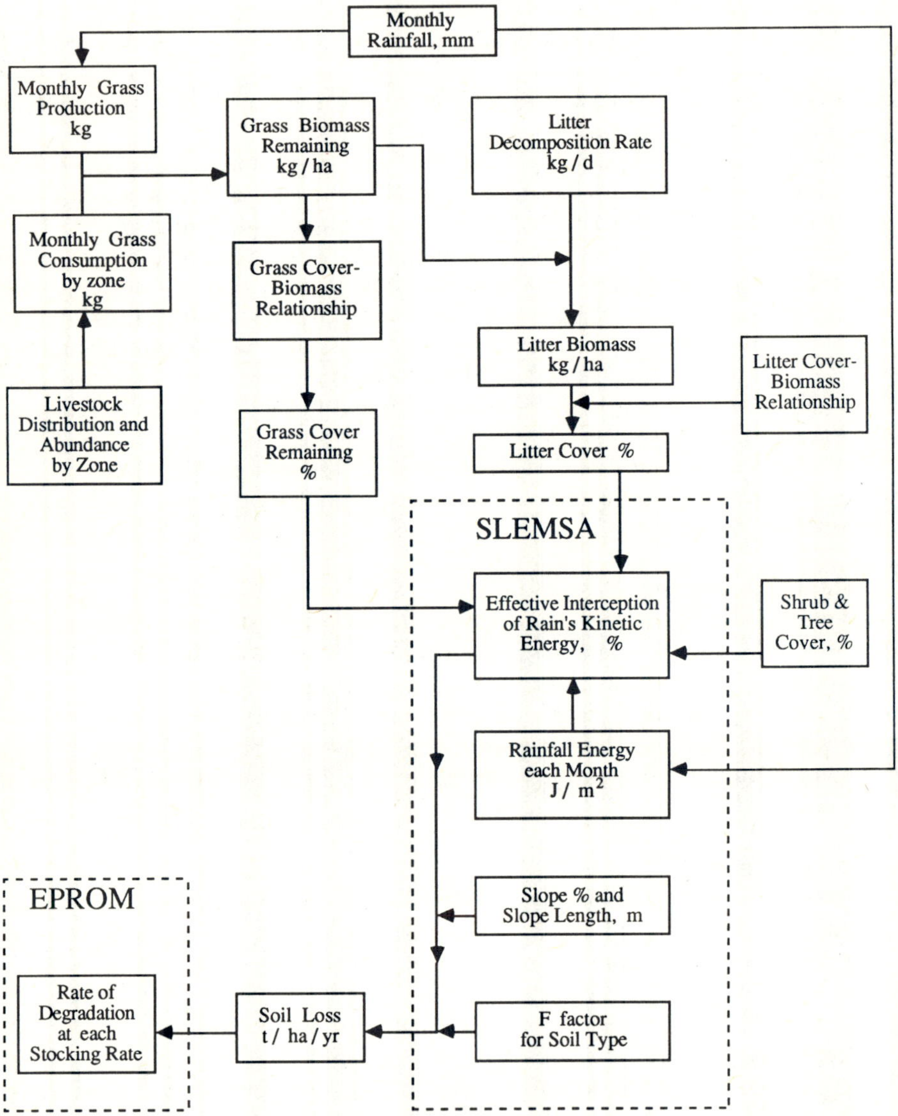

and biomass on the plots were then measured. There was a virtually unbiased and acceptably precise correlation between the visual estimates and measured cover, and a logarithmic curve relating estimated cover to measured biomass fitted well (Dancy et al., 1986). Estimates of standing crop were made by first estimating grass cover on the survey diapositives, then converting cover to biomass using the relationship established on the experimental plots.

The grass standing crop for May 1983, at the end of the growing season, was converted to annual production by estimating the quantities of grass that the estimated numbers of livestock would have eaten, and adding this to the observed standing crop. This procedure was carried out in the Pelotshetlha area in the South-east of Botswana for 1983 and 1984, and in the erosion model study area for 1983 (Fig. 11.1). When the annual production figures were plotted against rainfall, all fell close to the regression line for grass production on thornveld at Matopos in Zimbabwe (Dye and Spear, 1982). It was consequently assumed that the relationship between rainfall and production in the present study area was similar to the Matopos model. Dye's (1983) work was therefore used as a source for estimating monthly relationships between rainfall and grass production. A linear regression of production increment on the current and the previous month's rainfall proved satisfactory. This regression was used with monthly rainfall figures in the study area to estimate the evolution of grass over the rainy season. For the CSR, it was calibrated to produce the amount of grass observed in May 1983 for the pattern of monthly rainfall recorded. For the RSR, coefficients were adjusted to give a greater response to small amounts of rainfall but a similar response at 900 mm, in line with the findings of Kelly and Walker (1986) and McKay (1968) reported above.

Validation

The primary production, grass cover and consumption component of the erosion sub-model was validated in the study area of the Integrated Farming Pilot Project to the south, (Fig. 11.1), where Abel et al. (1987) had taken similar low-level aerial photographs and conducted surveys of livestock densities. The results of the validation are given in Table 11.2.

There is virtually no erosion by water in July and September because rain is scant. If the mean of the cover values for January and May are used to

Table 11.2: Validation of grass cover predictions

Month	*Grass Cover, %*	
	Predicted	*Measured*
January	59	54
May	54	61
July	50	57
September	45	50

estimate the rainy season cover profile, the mean cover predicted from estimated grass growth and consumption by livestock is 56.5%. The mean of the cover values for those months as estimated directly from aerial surveys is 57.5%. With a rainfall of 500 mm, a slope length of 30 m, slope steepness of 8% and F value of 3, the estimated cover value yields a gross soil loss estimate of 5.64 t/ha/y. The observed value produces a value of 5.61 t/ha/y.

Sensitivity analysis

The vegetation cover component of SLEMSA has a negative exponential form (Fig. 11.5). The rate of primary production was, as already noted, higher on average under the RSR than under the CSR to allow for past impacts of heavy grazing on soil. Varying either the rate of primary production or the stocking density can have the effect, in the more heavily stocked system, of reducing the cover to below the point of maximum curvature of the cover-soil loss curve. The shape of this curve has important implications for range management (Abel and Blaikie, 1989), as well as for the model. Above the point, the rate of soil loss is insensitive to the rate of either primary production or stocking; below the point, soil loss increases exponentially with an unfavourable change in either rate.

The other parameter in SLEMSA which has a strong influence on the rate of soil loss is the erodibility factor, F, the derivation of which is empirical. Biot (1988) derived values of F for a granite landscape by measurement, and these results guided the choice of values here.

Results and discussion

Fig. 11.3, and Figs. 11.6 to 11.12 show cattle densities, rainfall, energy output per ha and per Livestock Unit, gross margins per ha and per LU, and rates of soil loss. The rates of soil loss are for one soil type (on granite) on one land type (the hill zone), where mean slope was 8%. Rates of soil loss from other zones with much gentler slopes were very much lower.

Energy production and gross margin per LU were much greater at the lower stocking density (Figs. 11.8 and 11.10). This relative advantage is reversed in Fig. 11.9 and 11.11. which show that the CSR is much more productive than the RSR. Even in the depths of a drought (Fig. 11.6), the CSR produces more energy and a higher gross margin per hectare than the RSR.

The short-term advantages of the CSR as a means of supporting people are therefore clear, but what are the long-term consequences? Fig. 11.12 shows the relative rates of soil loss under each stocking rate. These are surprisingly similar, which can be explained by Fig. 11.5. Mean cover values for the eleven-year run were 51% for the CSR and 73% for the RSR. While substantially different, both values lie above the point of maximum curvature in the cover-soil loss curve, so that a large change in cover results in only a small change in soil loss. It should be remembered, though, that these results are from a model, and that if the effective cover of *both* systems proved to be in reality 20% lower than the estimate, so that the RSR system remained above but the

Figure 11.5: Gross soil loss and cover in relation to rainfall

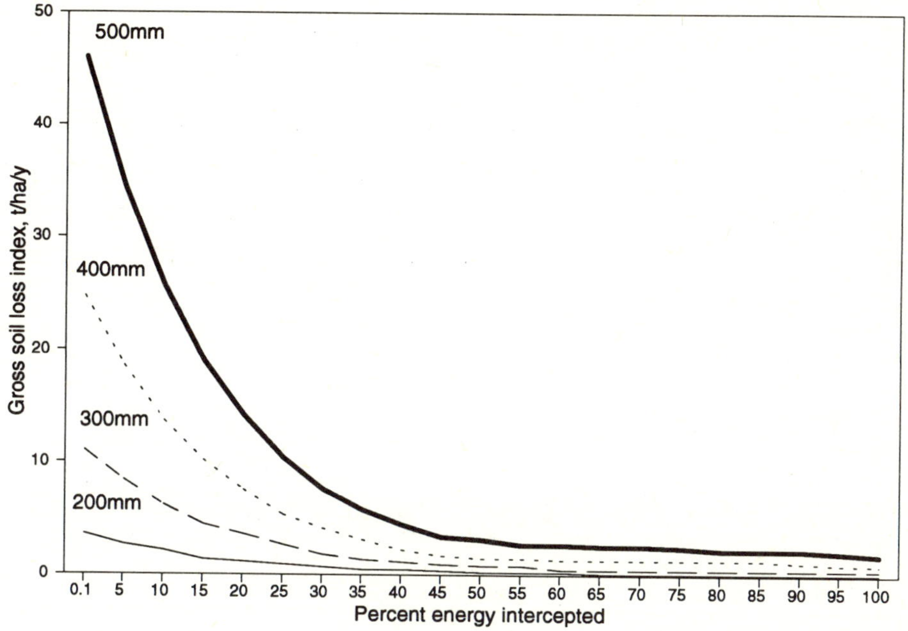

CSR moved below the point of maximum curvature in the cover-soil loss curve, the difference in rates of soil loss between the two systems would rise greatly. This aspect of the model was validated, however, and these cover values for CSR were supported thereby. Moreover, the rates of loss on other land types with gentler slopes were similar to those found by Biot (1988) on land with similar slopes. At such rates Biot predicted that the current rate of primary herbaceous production would be sufficient to support existing densities of livestock for around 400 years (see also Biot, in this book). The evidence from this model and from Biot's work therefore suggests that there is little to be gained from de-stocking.

Models are of course simplifications of reality, and it is necessary, in particular when a counter-intuitive conclusion emerges, to ask whether an important element has been omitted. This model indicates that the *direct* effect of de-stocking on soil loss is slight. It is possible, however, that a greater indirect effect could occur if grazing pressure initiated a change in species composition which much reduced effective cover, through the replacement of existing species by annuals, which afforded less cover. This is a different effect from the one modelled here, in which the effective cover is determined by the

Figure 11.6: Rainfall, Central Region, Botswana

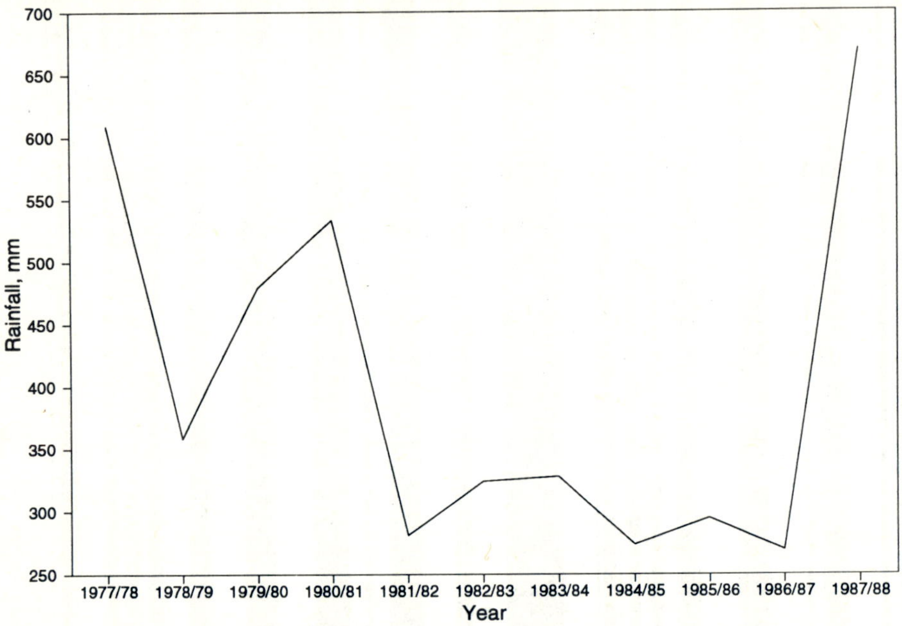

Figure 11.7: Rainfall in soil loss study area

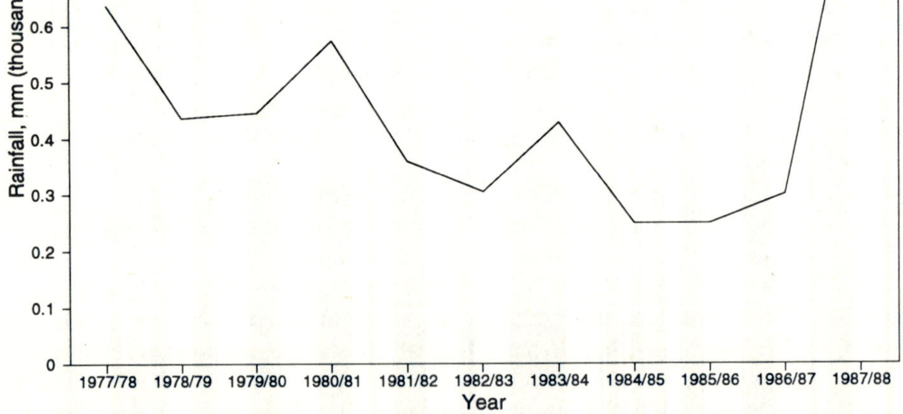

Figure 11.8: Energy output per livestock unit

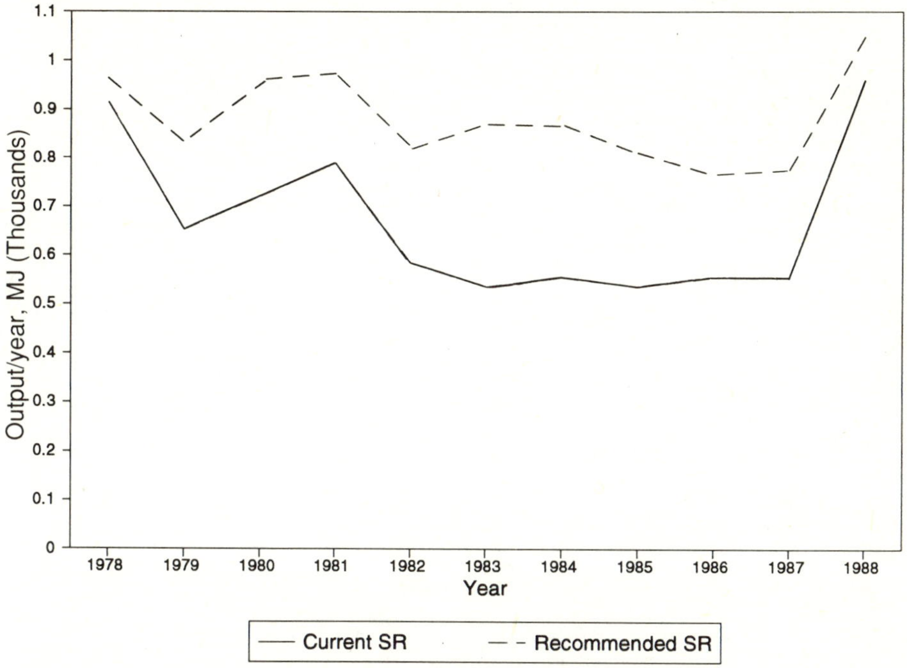

Figure 11.9: Energy output per hectare

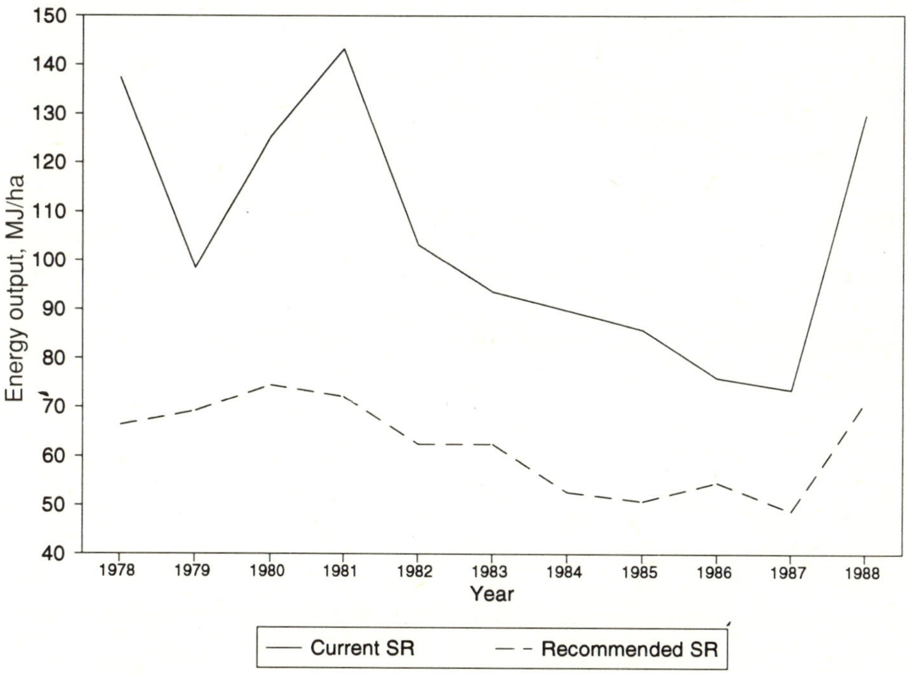

Figure 11.10: Gross margin per livestock unit

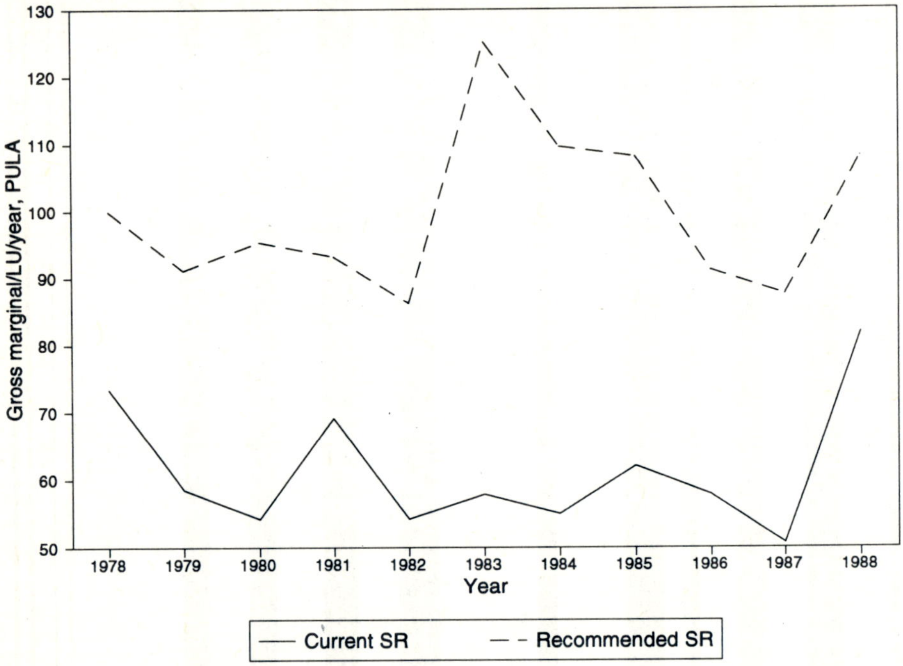

Figure 11.11: Gross margin per hectare

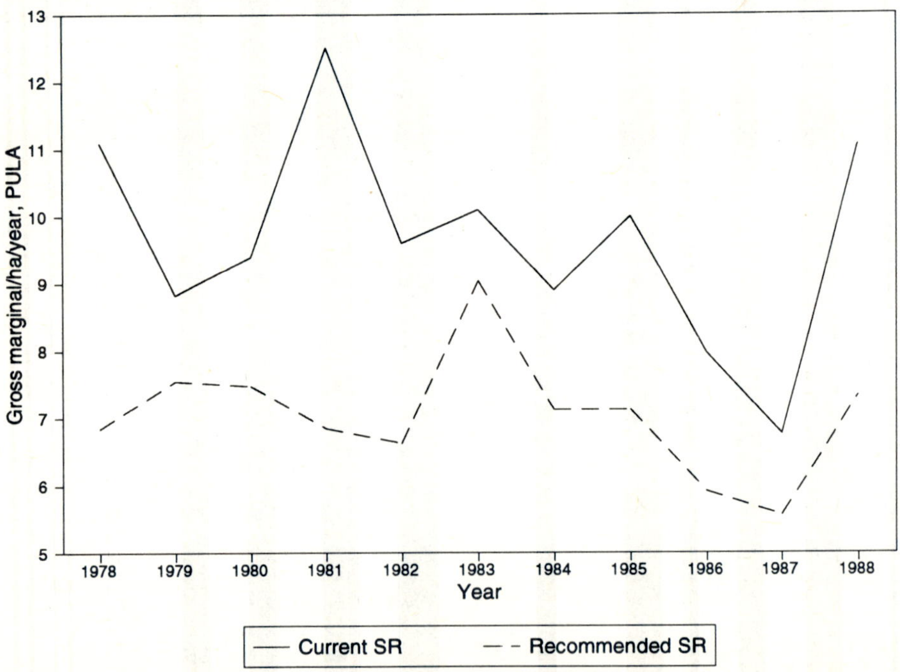

Figure 11.12: Soil loss at two stocking rates granite landscape, hill zone

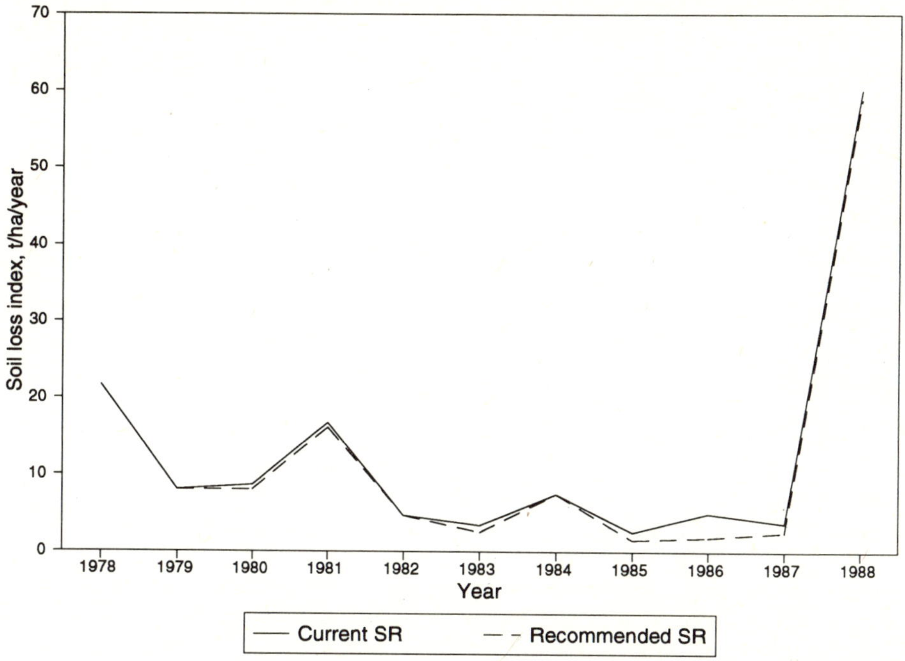

amount of grass removed by consumption. The species composition of the herbaceous layer is certainly very different in this study area from that on similar sites under research ranch conditions. A change has therefore already occurred, but this has not meant that cover has fallen below the point of maximum curvature on the cover-soil loss curve (Fig. 11.5).

Harrington et al. (1984) have suggested that most degradation of range occurs in the early stages of drought, when livestock density remains at pre-drought levels but the biomass of vegetation has greatly declined. Livingstone (1991) has rightly raised this issue in criticism of the 'average effective cover' approach taken by Abel and Blaikie (1989). He cites Kelly's (1973) measurement of grass production on moderately and on heavily grazed land in Zimbabwe, and estimates from Abel and Blaikie's (1989) simple model of soil loss from communal range, to suggest that in a dry year the rate of loss from the heavily stocked range may be seven times greater than that from the moderately stocked land. The principle is correct, but because rainfall is also low in a dry year, even the higher rate of loss will be a small amount of soil (Fig.11.5). This is illustrated by the relative rates of soil loss under low and high rainfall (Figs. 11.7 and 11.12).

The long-term effects of dense stocking have already been included in the model by making primary production more responsive to small amounts of rain under the RSR. In the CSR, at the beginning of the rains, grass growth is slower and cover more sparse, leading to greater soil loss. But for this, the difference in rates of soil loss between the two systems would be even smaller.

What the model does not do is to carry forward the negative effects of a particular drought, such as to reduce the rate of primary production when the rains return, and to carry forward a greater effect under the CSR than the RSR. This may be a desirable procedure, but no measurements of this process have been found in the literature.

A related limitation of the present model is the long time-step of one month imposed on it by data and computing constraints. If catastrophically high rainfall fell on drought-affected range, a large amount of soil could be lost in hours, before there was time for vegetation to cover the soil. In the present model, vegetation growth and soil loss are averaged over a month, with the risk of missing this effect. However, it is predicted that both lightly stocked and heavily grazed range would lose large amounts of soil when very intense rain follows drought, and that much more of the loss would be accounted for by the intensity of the rain than by the stocking density.

Conclusions and policy implications

This has been a preliminary presentation of a model to analyse some of the benefits and costs of de-stocking the communal areas of Botswana, with implications for policies on carrying capacity in Southern and sub-Saharan Africa. The results support the reluctance by pastoralists in the study area to de-stock. So long as production is limited by the area of pasture within grazing radius of water, and while the primary aims of cattle-keeping are security, savings and subsistence, it will remain logical for pastoralists to maintain high stocking densities. The logic will change if the production goals change.

The costs of dense stocking, in terms of rate of soil loss, appear to be fairly low, in that reducing densities drastically results in only a small decline in the rate of soil loss. In terms of Biot's model (chapter in this book), destocking would not extend the life of the soil significantly. However, the rate of loss is critically dependent on a threshold of vegetation cover; once below this, the rate of soil loss increases exponentially.

The findings here are applicable to Eastern Botswana, where most Botswana pastoralists live. Management practices in the study areas were typical of those in Eastern Botswana, but it is possible that stocking densities are higher elsewhere in the country: the spatially aggregated figures in the Agricultural Statistics do not permit the estimation of comparative densities. Thus it has not been established by this study that there are no other areas where densities are sufficiently high to drive grass cover below the point of maximum curvature on the cover-soil loss curve. Even if this was the case, effects on soil loss are unlikely to be as serious in other parts of the communal areas, as slopes are unusually steep in the main study area as compared to most of Eastern Botswana.

The methods expounded here are applicable to a wide variety of circumstances in other countries. The conclusions are not. The work presented here is not attempting to replace the belief that 'pastoral Africa is turning into desert' with the equally unscientific conclusion that 'there is no rangeland degradation'. The task for pastoralists and range scientists is to assess the rates and distribution of degradation, and then to determine, in cooperation, what should be done about it within each particular set of circumstances. Where land is not especially erodible, nor rainfall particularly erosive, resources should be put into matters of higher priority than the conservation of land that is not at significant risk.

12

Out of Africa, Looking In: Understanding Vegetation Change*

Mark Stafford Smith and Geoff Pickup

Introduction

While considerable progress has been made recently in developing process models of rangeland vegetation dynamics in response to climatic variability, grazing, fire and the various forms of land degradation, there is still no generally-agreed body of theory. This lack of a framework within which vegetation change might be interpreted has led rangeland science into a plethora of case studies, often showing apparently contradictory results, and has created disillusionment and scepticism among both practitioners and clients, problems expressed eloquently by Mentis (1985) and Mentis et al. (1989). The failure of rangeland science to produce such a general model or framework has meant that it is extremely difficult to assess whether particular land uses or grazing management approaches are sustainable within a particular time frame, or whether they are continuously or intermittently degrading the soil and vegetation resources upon which they rely (Pickup and Stafford Smith, 1993). Specific problems in assessing sustainability include the difficulty of separating grazing-induced degradation (or long-term loss of productivity) from short-term grazing effects and from natural change (Pickup, 1989) and the problem of monitoring changes in vegetation in an efficient and meaningful way (cf. O'Reagain and Mentis, 1990).

The conceptual model of vegetation change in rangelands has rested on the Clementsian succession model (Stoddart et al., 1975) for over half a century. The adequacy of this model in non-equilibrium systems is now being challenged, and new approaches developed, expressed as 'state and transition' models (Westoby et al., 1989).

These new approaches to rangeland dynamics improve on the earlier models by being more holistic and realistic, but they lack any spatial component. Rangeland managers must deal with vast areas of heterogeneous landscape where the linkages and flows *between* parts of the landscape may submerge or even reverse the type of plant behaviour that might be expected at individual points from a state and transition model. Do the new approaches therefore provide the most appropriate information for the management of the vast rangeland regions with erratic rainfall?

---

* We are grateful to Roy Behnke and Ian Scoones for precipitating this review of our concepts.

During the last five years, there have been significant advances in our knowledge of the processes leading to spatial and temporal variability in Australian rangelands. Much of that knowledge has come via the analysis of multi-temporal satellite imagery in which patterns of change could be observed, expressed and modelled in relative terms (e.g. Pickup and Chewings, 1986, 1988a, b). Steady progress has also been made in ground-based studies (Stafford Smith, 1988; Tongway and Ludwig, 1990) but the logistics of making labour-intensive measurements at the many sites necessary to represent a spatially-variable process remains a limiting factor.

The problem lies in the complexity of rangelands: many plant species interact, large areas are involved and processes vary dramatically through time and space. In this chapter, we begin by reviewing approaches to process modelling in rangelands. The limitations of these models lead us to show how variability in time and space influences vegetation dynamics, and to argue that it is the slower changes and spatial fluxes which are most poorly understood. We illustrate how spatial patterns in landscape features, in the supply of resources and in all driving forces make up a set of building blocks which can be used to generate complex spatio-temporal patterns of behaviour. Given this complexity, the need for conceptual simplification becomes evident, and we discuss the adequacy of various paradigms of vegetation dynamics in the context of providing appropriate information for management and policy. We finish by discussing whether the ideas developed in Australia can be transferred to other continents.

Processes in vegetation change

Like rangeland systems in many parts of the world, the natural physical and biological system on which Australia's rangeland industry is based is highly variable in space and time (Stafford Smith and Morton, 1990; Pickup and Stafford Smith, 1993). Here we consider what the important parameters of the system are and how they affect behaviour through time at a single point; we then look at how existing process models reproduce this behaviour.

Resources, system parameters and driving forces.

The establishment and growth of vegetation at a point in the landscape occurs in response to the availability of seed, and three resources: water, nutrients and energy (taken here to be in the form of solar radiation). These resources vary as follows:

(i) seed may be available from a local source, or by redistribution from elsewhere (termed the *mass effect* by Shmida and Ellner, 1984); its survival may be affected by climatic conditions and predation.

(ii) water availability depends on the balance between infiltration and storage in the soil, and evapotranspiration; water may be supplied directly from rainfall, but that supply can be greatly reduced or increased by the redistribution processes of run-on and run-off.

(iii) nutrient availability likewise depends on the balance between the rates of turnover in relation to the total stored in soil and vegetation, and the competitive use of the resource by other plants; nutrients can also be redistributed across the landscape by wind, water or animals.

(iv) solar radiation input is largely pre-determined by latitude, but it can vary locally with topographic factors such as slope, aspect, and shading by topography or other plants.

Vegetation change at a particular point in the landscape may take place when there is a change in the availability of any of these resources or in the intensity of biomass-removing processes such as grazing and fire. Thus vegetation change results from the interaction of existing conditions or *system parameters* at a point and certain imposed *driving forces*. Key examples of these for most rangelands are shown in Table 12.1, although other factors are important in some areas of the world. Climate is outside the control of management; the other driving forces can be modified by management, and these modifications may determine how the effects of climate are manifested. It is useful, therefore, to consider management effects *given* the constraints of climate.

Change in vegetation at a point can be regarded simplistically as a process of rain and net run-off resulting in a store of soil moisture which, above certain thresholds, is used for plant growth subject to constraints imposed by nutrient supplies and temperature. Different plant species potentially compete for the water, nutrient and perhaps light resources. Animals (or fires) then remove certain portions of the plant biomass which affects subsequent plant growth. The animals may grow, reproduce, and die or be sold, resulting in economic turn-off for the enterprise. Naturally, this brief description hides many complexities, especially in the ways in which system parameters change. Before examining these complexities, we briefly review how existing process models approach these problems.

Process models of vegetation change

In Australia, as elsewhere, most research effort has gone into developing models of short-term vegetation change in relation to rainfall, since these measures of plant production have obvious economic application. For example, the GRASP model of McKeon et al. (1982, 1990) takes the most common approach to production based on a soil moisture budget and plant water use efficiency. It contains various assumptions specific to the grassland environments in which it was originally developed, but now has the advantage of a large test database of production data for many environments in Queensland, so that it is now being applied on a 10 km grid across the whole state (G. McKeon pers. comm.). It has been modified to account for nitrogen limitations in sub-tropical grasslands with their differential effects on C3 and C4 grasses, and, in a simple way, for germination in summer-rainfall arid environments. The model also contains well-tested empirical relationships which describe the interactions between grasses and a tree overstorey (Scanlan and Burrows, 1990), although the theoretical basis for these is not clear. Other

Table 12.1: Key resources, system parameters, and driving forces of importance in the Australian rangelands

Resources
- water
- nutrients
- energy (solar radiation)

System parameters ('response functions')
- infiltration rates through the soil surface
- soil column water storage capacity, and current status
- soil and vegetation net available nutrient pool
- water and nutrient use of existing living vegetation
- flammability and palatability of existing living vegetation
- the suite of species (with their endogenous characterisics, such as growth responses to temperature) available to colonise the site
- position in the landscape through its effect on radiation, and water, wind and seed redistribution

Driving forces ('forcing functions')
- climate, in particular rainfall and temperature, which affect the local input and turnover rates of water and nutrients
- grazing, by domestic and unmanaged herbivores, and fire, both of which affect the rate of removal of above-ground biomass
- processes which redistribute water, soil, nutrients and seeds across the landscape, thereby changing resource availability at a point

vegetation models in Australia include GrassGro (Donnelly, 1988) which deals with temperate grasslands, but is being extended to handle mixed-species pastures in other areas; the model of Noble (1975) which describes five plant categories (two shrub species, short-lived grasses, forbs and standing dead material) in a chenopod shrubland and includes germination effects for the short-lived species; and IMAGES (Hacker et al., 1991) which contains shrub growth relationships within its control theory approach to a whole property. All of these production models work satisfactorily in the appropriate environments, and compare well with models developed for other continents (e.g. Hanson et al., 1988; Seligman and van Keulen, 1989; Stuth et al., 1990). However, all deal with, at most, a few key species, and often collapse all production into one or two classes of vegetation. Most do not handle establishment of longer-lived plants, and consequently are restricted to the short-term dynamics of forage production in response to climate.

Longer-term vegetation change is handled separately in a few other approaches. SHRUBKILL (Ludwig, 1990) includes functions which determine how the use or lack of fire may impinge on grass-shrub balances in the pastoral regions of New South Wales. The vital attributes model of Noble and Slatyer (1980) aims to predict the coarse interactions between plants with different life history strategies, and early formulations which predicted only the presence or absence of species have been replaced by a more sophisticated approach which includes some competitive interactions (Moore, 1990). No comprehensive application has yet been made in the rangelands but examples have been drawn from there. Some specific interactions have been modelled more completely elsewhere; e.g. mesquite-grass balance in Texas (Archer, 1989), long-term soil nutrient dynamics in Parton et al., (1989), and various other studies usefully straddle the boundary between process modelling and conceptual simplifications (e.g. Noy-Meir, 1975; Walker et al., 1981; Noy-Meir and Walker, 1986; Tilman, 1988). In general, however, there is a substantial gulf between approaches to long-term vegetation dynamics and the more process-driven, short-term production models.

For vegetation production to be of relevance to managers, it must be possible to grow animals from it and to assess the economic implications of various turn-off strategies so the feed-backs of animals on the vegetation can be determined. Given forage production and quality, as well as total offtake at the particular point, animal production is quite well understood. The most notable models, which draw on approaches developed outside the rangelands, are the PROBE model of McKeon and Rickert (1984), and GrazFeed which implements the Australian feeds standards (Standing Committee on Agriculture, 1990). Again, good models are also available in the United States, as part of the models already mentioned. The problem is not so much one of predicting animal production but rather of knowing the quality or palatability of future plant production if vegetation composition is liable to change, and the offtake at the point given that grazing is not uniform in space.

Finally, biological production in rangelands needs an economic interpretation. Models which integrate herd dynamics with property economics vary from spreadsheet programmes (Stevens and Peggs, 1988; Holmes, 1988) to more flexible stand-alone systems (Taylor et al., 1987; Stafford Smith and Foran, 1990). Few of these handle variability through time very well, although risk has been addressed in ways that managers could follow (Foran and Stafford Smith, 1991; Stafford Smith and Foran, 1992). Outside of Australia, stochastic dynamic programming has been used for similar purposes (e.g. Rodriguez and Jameson, 1988; Mace and Houston, 1989). None of these approaches deal explicitly with the feedbacks between forage consumption and future plant behaviour. There is a vast body of economic theory available (e.g. Dillon and Anderson, 1990) but relatively little of this has been integrated into rangeland models (e.g. Stuth et al., 1990).

Most of the models described above deal with production as if it were a point process. Spatial variability which is fixed in time may be incorporated into models of vegetation dynamics simply by dividing the area of interest into different types and applying a separate point or landscape-average model to each in order to obtain a set of representative or mean responses to a particular perturbation (e.g. Burke et al., 1990). Some patterns of spatial variability may change over time in a random or systematic way independently of the vegetation and can be incorporated in vegetation models as space and time-variant behaviour, without feedback, using appropriate statistical processes (e.g. Noy-Meir, 1973). An example of such an effect might be the differences in water input on a sand plain without run-off as a result of spatially-stochastic local storm activity. However, other types of spatial variability are functions of vegetation state or response and have to be represented by explicit flows which result in feedback loops in the model.

There are few examples of models which deal with spatial variation in any of these ways. The simplest approach to handling a fixed spatial pattern is to identify an apparently critical component and ignore the rest: GRASP has been used in this way to predict safe production levels off heterogeneous paddocks (G. McKeon pers. comm.) by considering only the most productive or most easily-damaged landscape unit. The next level of sophistication is to break the management unit down into supposedly homogeneous components and to model each separately. The first example of this was Goodall (1967), who modelled a paddock in several compartments but basically ignored flows between the components.

Spatial flows and feedbacks across flat rangeland landscapes were allowed for by Noble (1975) and Stafford Smith (1984, 1988) who broke a paddock up into a regular grid and simulated water flow and animal movement between grid cells, as well as plant growth and offtake as point models in each cell. This approach worked but was clumsy, computationally intensive and undoubtedly overextended our current understanding of animal movement and selection behaviour. Soon after this Pickup and Chewings (1986) developed stochastic auto-regressive spatial models of erosion patterns, and Stafford Smith (1988) used simpler regression techniques to characterise grazing distributions. There seem to be no other examples of such spatial models in Australia, and few elsewhere (but see Senft et al., 1983; Namken and Stuth, 1991; Coughenour, 1991), despite considerable concern about spatial patterns in various contexts (e.g. Harris and O'Connor, 1980; Senft et al., 1987; Andrew, 1988; Schlesinger et al., 1989; Behnke and Scoones in this book).

In summary, existing models of vegetation dynamics in rangelands handle short-term plant production at a point quite well and can be linked with appropriate animal and economic production models. However, the models fail to deal adequately with long-term vegetation change or with the impact of spatial redistribution. The remainder of this chapter seeks to demonstrate that these missing components can result in damaging limitations to the existing models and that alternative approaches can provide more efficient ways to tackle the problem of predicting vegetation change.

The nature of change through time

Given the limitations of the process model approaches reviewed above, we now step back to examine some limitations of the traditional framework of process understanding.

Scales and rates of change

A component of a vegetation response model may vary through time or over space at a range of scales; whether this variation is considered to be static, slow or fast depends on its rate of change compared with that of the plant community. For example, biomass changes rapidly in arid areas in time periods of the order of weeks; vegetation composition changes less rapidly, although the rate is partly dependent on the longevity of dominant species; meanwhile, vegetation structure usually changes only slowly, and the pool of potential colonising species is effectively static in the absence of introductions. Thus there is a continuum from static to dynamic but with endpoints that vary with the plant attribute of interest. In pastoral management, our object is usually to feed and to keep on feeding grazing animals. We are therefore interested in both the quantity and quality of forage produced by rainfall, which involves fast processes, and in the ability of the system to keep on producing that forage, which changes at a slower rate.

Given a sequence of events such as rainfalls, the fast processes which determine vegetation response will produce an associated pattern of vegetation states which can be characterised in terms of its mean, variance and autocorrelation structure or 'memory'. Memory results from long-term impacts or lags, such as changes in soil condition or vegetation composition which may persist long after the causative event; most systems have some memory, and an increase in that memory results in more continuity in responses through time. Many of the differences in behaviour between sites can be expressed as a combination of these building blocks. Furthermore, the pattern of vegetation states through time can change in relation to any or all of these factors, and the changes may be gradual or stepped (Fig. 12.1).

The fast processes involved in forage production at a point are reasonably well understood, and amenable to modelling using process-based approaches (see previous section). Put simply, long-term vegetation change is dependent on the same processes, given differential responses to the driving forces by different plant species. However, this is not an efficient or even practicable way of determining vegetation change when a hundred or more species may be involved. The slower processes, which change the ability of a system to produce forage, involve changes in factors such as soil moisture storage capacity, nutrient status and the availability of a seed bank. These processes operate at a rate which is comparable to a manager's working life and on a large spatial scale which makes them difficult to observe, measure or comprehend. Consequently, slow processes are poorly understood. We now consider how climatic variability and the other driving forces affect these processes.

The role of climatic variability

Climate influences soil properties, nutrient recycling and vegetation growth in numerous ways but these effects are dominated in arid regions by the availability of soil water. Australian rangelands encompass areas with everything from summer- to winter-dominated rainfall but the rainfall regime is always *variable* and *unpredictable* (Nicholls and Wong, 1990; Stafford Smith and Morton, 1990). Unpredictability of rainfall is as important as variability because it reduces the number of management options available. From a biological perspective, rainfall records may be treated as *events* and *sequences*

Figure 12.1: Some of the ways in which responses can change over time

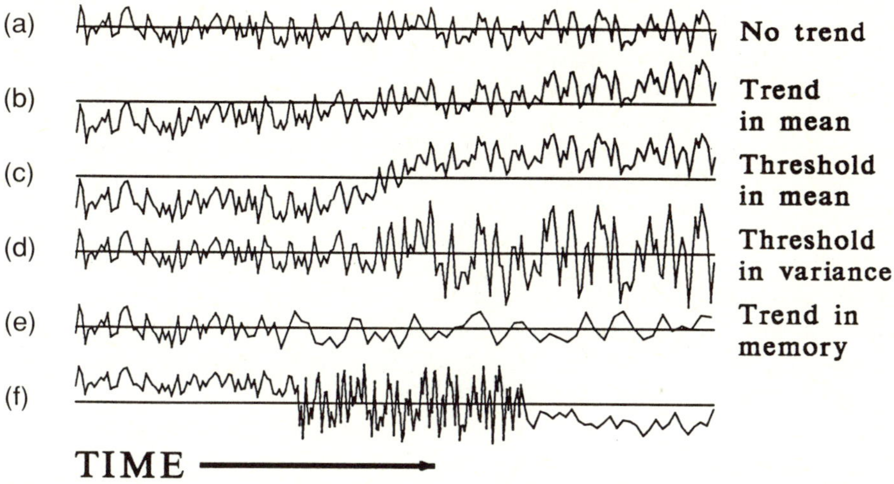

Notes: In each case the y-axis represents a measure of state while the x axis is time, for example, (a) seasonal fluctuations in cover on stable landscapes, (b) succession in woody biomass (this may also involve (e)), (c) vegetation cover change in response to fire or seasonal changes, (d) aggregate vegetation response to rainfall in a degrading landscape, and (e) an increase in system memory, for example during a change from short-lived to long-lived plant species. Changes in variance or auto-correlation (or memory) may be gradual or stepped, as illustrated for change in means in (b) and (c). (f) illustrates a more complex change in level of primary productivity for grazing, where a perennial grassland (high mean, small variance, moderate memory) is altered to an ephemeral herbland (moderate mean, large variance, very little memory) which is then encroached by unpalatable shrubs (low mean, small variance and considerable memory) (cf. Westoby et al., 1989).

---

of events. An event is one or more closely spaced rainfalls (or soil moisture impulses) which are large enough to produce a significant vegetation response - a substantial pulse of plant growth which produces forage for livestock. This translates, for example, into events of about 50 mm in a week in the Alice Springs region of Australia. Several events may occur within one year, but it is also possible for several years to pass without any (Stafford Smith and Morton, 1990).

Individual rainfall events determine short-term variability in the biological system: for example, the presence or absence of ephemeral plants and the growth periods of perennials. However, they are not usually large or persistent enough to change the system's ability to respond to future rainfall. Sequences of events *are* capable of producing natural change in slower processes: for example, by diminishing the seed bank of certain plants over time, or establishing dense thickets of woody shrubs. The sequences consist of a run of events (and periods without rainfall) of sufficient length and magnitude that short-term changes in the biological system are progressively reinforced until a new state is reached (Westoby et al., 1989; Friedel, 1990). In this circumstance, for example, a species may become locally extinct or dominant. Changes may occur gradually, or as one or more discontinuous steps as in the 'catastrophic sequences' model of Griffin and Friedel (1985), although the difference between these really depends on the timescale of interest.

The impact of a sequence of rainfall events usually overrides other factors in causing short to medium-term variability in both production and composition of vegetation (especially for short-lived plants). However, the effects expected from a rainfall sequence may be damped, enhanced or even reversed when the impact of processes such as grazing and run-off redistribution are added. It thus becomes possible to have quite different effects from the same event in adjacent areas (Pickup and Chewings, 1992). The critical requirement then becomes one of how to detect trends due to driving forces other than climate through all this noise. In the next section we examine what form those trends might take.

Grazing and fire

Grazing offtake and fire both share the property of removing above-ground biomass and of causing plant mortality in some cases. Both forces may vary in intensity through time, although fire is usually more episodic and associated with times of above-average rainfall. Both forces are selective in their impact: grazing selects against palatable plants whilst fire selects for fire-adapted species. Plants can invest resources in making themselves less liable to impact from either force. The two forces certainly interact, in that grazing can suppress fuel for subsequent fires, and, less significantly, fire can remove fodder from herbivores. Both forces can also have indirect effects on vegetation, through trampling impact on soil properties in grazing (e.g. Harrington et al., 1984; Tongway, 1990), and through losses to the nutrient pool as a result of fire (Harrington et al., 1984; J. Walker et al., 1986). Grazing and fire can both be managed, so our key need is to understand how these forces may modulate the effects of climatic variability.

The removal of biomass is a fast effect - singular in the case of fire, and chronic with grazing. The results of biomass removal are generally straightforward to predict in the short-term, since it merely alters forage availability. Reduction in cover can cause short-term changes in water use by plants, in infiltration, and in protection from erosion. Grazing may have more subtle effects on seed production by plants since this is usually the first growth function to suffer when plants are put under stress. Some species of grasses in New South Wales, Australia, for example, have been shown to cease effective seed production at levels of grazing two to three times lower than those at which significant plant mortality is noted (Hodgkinson, 1991; for a counter example, Milton and Dean, 1990).

Grazing selection and soil changes have slow effects on processes affecting vegetation composition. These are harder to predict because of interactions between species and the need to have a detailed knowledge of the properties of different species in relation to palatability, flammability and fire tolerance.

Grazing has extra complexity where several herbivores with their different preferences are concerned; fire response similarly may be very dependent on season and intensity of fire. Some means of simplifying and grouping species (e.g. by *vital attributes*, Noble and Slatyer, 1980; Moore, 1990) is therefore needed in all but the simplest vegetation types. Furthermore, there may be feedbacks; these may be negative, for example when grazing causes loss of palatable species and consequent lowered use, or positive, such as when grazing keeps biomass young and thus enhances patch grazing (Andrew, 1986). The impacts of grazing may occur continuously, as with the effect on seed production mentioned above, or at critical times, during droughts or during new shoot growth (e.g. Mott et al., 1992).

Changes caused by grazing affect not only the site parameters such as soil infiltration and moisture storage capacity. They also influence the temporal pattern of plant production. For example, a shift from palatable perennial grasses to ephemeral vegetation results in a great increase in variability (Friedel et al., 1993), whilst a subsequent expansion of unpalatable perennials may reduce this variability (Fig. 12.1f). These changes are important to management, since it is easier to cope with risk in systems with low variability. Whether the system contains palatable or unpalatable species also obviously determines the stable production level.

The nature of change over space

The basic problem in considering vegetation dynamics only through time is that changes in resource levels do not result from point processes: increases in one part of the landscape usually reflect decreases in other parts, and there may be critical feedbacks between the areas. Consequently, changes through time interact intimately with changes over space, to which we now turn.

Differences in vegetation response at the landscape scale may arise because of spatial variability in edaphic factors such as the soil environment, because of redistribution of water, soil, seed and nutrients, and because of spatial variability in the driving forces of climate, grazing intensity and fire. Behaviour involving feedback links between different parts of the landscape - such as the linkage between the intensity of water and sediment redistribution in erosion cell mosaics and the amount of plant cover (Pickup, 1985; Stafford Smith and Pickup, 1990) - is the least understood and the most difficult to incorporate in models of vegetation dynamics. In general, though, the spatial flows of resources must be explicitly recognised to make sense of the feedbacks, and whole landscapes must be modelled simultaneously. Whilst this can be attempted by linking vast numbers of point models together, there are more efficient ways of extracting the critical factors, some of which we outline in this section.

Static spatial patterning

Most geomorphic and topographic spatial patterns change very slowly and can be regarded as fixed in time or static (see Plate 12.2). Many landscapes are the product of past climates and the processes which created them are no longer active. Other landscapes are less easy to characterise and may contain a variety of surficial landforms, each with a different age and frequency of disturbance by sedimentation processes. For example, in a study of event frequency and landscape stability on the floodplains of central Australia, Pickup (1991) has identified deposits emplaced by superfloods which may have changed little in the last 1,000 years. Adjacent to those deposits are narrow threads of sediment which make up the currently-active floodplain and which are disturbed almost every year. Plant behaviour on these surfaces differs markedly (Friedel et al., 1993) because of the associated spatial variability in soil moisture-capturing abilities as well as water supply.

Few maps of landscape attributes make the distinction between static and dynamic spatial variability although the difference can sometimes be inferred. In contrast with parts of Africa, Australian rangelands are well-mapped into land systems and sometimes land units (Christian and Stewart, 1968). Unfortunately, most land systems contain too much internal variability to provide a basis for vegetation response modelling. Land units offer more promise but, because they are an amalgam of topography, pedology, hydrology and vegetation, they may reflect long-term geomorphic history rather than contemporary patterns and processes of change. A new approach to landscape classification is needed for rangeland management in which factors which determine the risk, magnitude and potential direction of change in vegetation characteristics are the variables to be mapped.

Fast and slow processes in resource redistribution

Moisture and, to a lesser extent, nutrients are the principal limiting factors to plant growth in arid and semi-arid rangeland environments. While moisture, in the form of rainfall, may be delivered fairly uniformly over time across a landscape, it is redistributed by the processes of run-off and run-on. This means that moisture supply and associated plant growth opportunities not only vary with storm size, they can also vary in space by more than one order of magnitude. Redistribution processes therefore provide a mechanism for modifying vegetation response to grazing to the extent that totally different behaviours may occur in adjacent parts of the landscape (e.g. Pickup and Chewings, 1993). Students of plant behaviour who, of necessity, work with quadrats or small plots tend to have to treat this type of variability as problematic noise in their data rather than as a fundamental landscape property. However, a knowledge of redistribution is especially important in situations where changes in plant cover affect the amount and spatial pattern of run-off, creating a feedback loop and imposing a particular trajectory on plant community behaviour through time (e.g. Pickup, 1988; Tongway, 1990; Stafford Smith and Pickup, 1990). Where these feedbacks occur, plant behaviour is increasingly governed by spatial processes and cannot be explained without them.

The most important form of redistribution in rangelands involves water, not only because plant growth is moisture-limited, but also because of what water transports. Surface flow carries soil, organic material including seed, dissolved and sediment-bound nutrients and, sometimes, material which is toxic to plants. Sub-surface flow is more important for transporting dissolved material but the moisture redistribution pattern it creates may be particularly important for deeply rooted trees and shrubs.

Aeolian transport of sediment and organic material is a less important agent of redistribution since it does not involve the primary growth limiting factor but it can enhance the spatial patterns generated by water. For example, areas with little plant cover tend to shed dust and organic material, while areas with more extensive plant cover modify both air velocity and turbulence which promotes aeolian deposition. Little is know about the rates at which these processes operate but they are certainly important in the long-term and may also be very significant for upslope redistribution of seeds. It is also worth noting that a significant part of the topsoil in some parts of Australia's rangelands is a Quaternary dust mantle (MacTainsh, 1989). For the present, however, we concentrate on redistribution of and by water.

The patterns and processes of water redistribution (and the techniques to analyse them) differ with the amount of topographic variation. In steeper landscapes with organised drainage networks, flow paths are topographically constrained and slope is the most important hydraulic variable. In flat rangelands, where overland flow is the main run-off process, hydraulic roughness becomes the most important parameter in the flow equations. Hydraulic roughness is determined principally by the amount and type of vegetation cover, so an important feedback exists in which the spatial pattern of plant cover largely determines the spatial pattern of run-off and run-on. Plant communities therefore generate their own soil moisture regime within the constraints imposed by the climate and, initially, by the physical properties of the soil. In the longer term, even soil properties can be changed because both erosion and deposition rates are closely related to the amount of plant cover present. Almost invariably, eroding areas have less nutrients, lower infiltration and a smaller moisture-holding capacity than depositional areas (Tongway and Ludwig, 1990; Miles and Johnston, 1990).

Rangeland areas with significant topographic variation usually show a high level of water and soil redistribution but the constraints imposed by topography make the resultant patterns relatively easy to predict (Plate 12.1); these have been studied extensively (e.g. Lane and Nearing, 1989; Kirkby et al., 1990; Thornes, 1990). Variations in slope also influence the rate of sediment transport and the physical characteristics of the material in transit (especially grain size) which, in turn, determines the properties of the soils created when that material is deposited. Topographic effects remain more or less constant over time and can be treated as static. The pattern of redistribution, however, varies with the magnitude of the run-off event and the amount of plant cover present so it may be different when run-off occurs during a drought or in a wet period (Plate 12.1).

There has been little research progress until recently on flat rangelands. Now, however, an increasingly mathematical theory is developing which explains how plants and soil and water redistribution processes interact to produce a particular type of landscape behaviour (Pickup and Chewings, 1986, 1988a; Pickup, 1988; Stafford Smith and Pickup, 1990). The theory assumes that each landscape has a certain potential for redistribution which may or may not be realised, depending on the extent to which land is degraded. Some of that potential depends on the geomorphic history of the landscape. For example, at one extreme, aeolian-derived dunefields and sand sheets have very little potential for water redistribution because of their flat nature and high infiltration into the sand. This means that, even when most of the vegetation cover has been removed, there is virtually no increase in run-off. Spatial patterns therefore remain largely static or change in response to wind and spatially random events such as fire. At the other extreme, piedmont footslopes, alluvial fans and riverine desert may have clay-rich or duplex soils on which run-off is greatly reduced by the presence of vegetation cover. As these landscapes are defoliated, run-off and associated erosion increase. Eroded surfaces often become scalded or sealed, reducing infiltration and with it, the number of opportunities for plant germination and recruitment, thereby further increasing the potential for erosion (Pickup, 1985). If there is no organised drainage network, run-off and sediment tend to accumulate in hollows in the landscape where plant growth is enhanced, leading to further deposition. Redistribution therefore becomes a self-amplifying process because of these feedbacks and may reactivate relict patterns of erosion and distribution in the landscape or lead to the development of erosion cell mosaics.

Erosion cell mosaics are made up of a set of interlocking and overlaid erosion cells of different sizes and intensities. Each erosion cell consists of a source or production zone, a transfer zone and a sink (Fig. 12.2). The evolution of erosion cells through time is described by Pickup (1988). As the landscape becomes more eroded, source zones intensify and expand, transfer zones shift towards erosion and the larger sinks accumulate more material. Larger erosion cells intensify at the expense of smaller ones and the landscape is increasingly partitioned into areas of bare ground and heavily vegetated sink (see Fig. 12.2). Grazing can also cause some redistribution of nutrient resources, as is well-known outside arid areas (e.g. Taylor et al., 1985) and is visible as high levels of plant growth in intensively-used areas (see next section) such as waterpoints and camp sites in rangelands. These areas may be associated with weed infestations. However, this effect is probably not substantial at the scale of the whole landscape and we do not pursue it further.

Patterns of grazing and fire

The effects of grazing and fire occur in partly predictable patterns on the landscape. Future patterns are also somewhat dependent on past patterns, so that there is significant landscape-scale feedback. Grazing can cause short-term or permanent change in vegetation patterns which will alter grazing in the near or longer-term future. Fires will not recur immediately where a previous fire has removed fuel, but more significantly landscapes with an historically patchy pattern of burns are likely to retain the patchiness. Fire in general is more important as a management tool than as a natural force on most Australian pastoral rangelands today. As a management tool, its rates of spread, resulting patterns and likely impact in relation to subsequent climatic

Figure 12.2: The concept of an erosion cell mosaic (after Pickup 1985, 1991)

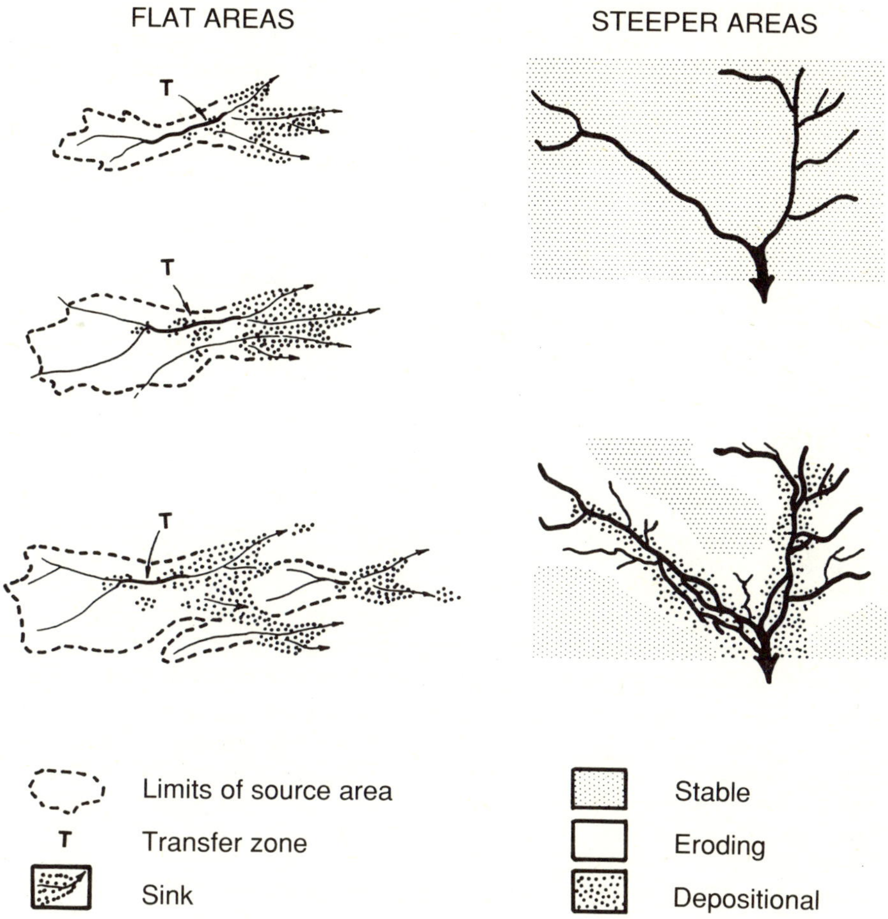

Notes: Source zones are areas of erosion and discontinuously shed eroded soil, organic material and run-off. They also have a low infiltration capacity which inhibits plant growth. Transfer zones occur downslope from source zones and are characterised by alternating erosion and deposition. On average, they are poised between erosion and deposition but can be tipped either way depending on the amount of vegetation cover present. Sinks accumulate the material transported from other parts of the erosion cell and are often occupied by dense woody vegetation because of the extra soil, water and nutrients they receive. Not all zones of each erosion cell occur in an erosion cell mosaic because larger cells may capture smaller ones. In flat country cell shapes vary and may appear random. In slightly steeper country, topographic effects may influence the pattern but slope remains less important than the vegetation growth-erosion/deposition feedback.

conditions are quite well understood (e.g. Griffin and Friedel, 1984; Hodgkinson and Harrington, 1985; Morton and Andrew, 1987; Noble, 1989; Braithwaite, 1991).

Grazing activity in arid and semi-arid rangelands is centred on artificial and natural waters to which animals must return at regular intervals to drink. Grazing intensity decreases with distance from water, and the distance animals are able to travel to grazing areas has physiological limits which vary with breed, herd structure, the salt content of the water and the herbage and temperature (Stafford Smith, 1988). Within these physiological limits, there is considerable spatial variability in the pattern of grazing largely as a result of differences in forage palatability. Indeed, diet preference factors are so strong that cattle, for example, will by-pass less palatable forage close to water and trek 10-12 km to favoured grazing areas. The resulting patterns can be complex, and vary through time (see Plate 12.2A).

Grazing intensity can be a major factor influencing the amount and species composition of vegetation which grows in response to a rainfall pulse. Spatial patterns of grazing activity must therefore be built into landscape-scale models of vegetation behaviour. At the same time, the quantity and quality of forage available at different locations affect the spatial pattern of grazing, so there is an important feedback mechanism. At present it is still unclear to what extent freely-ranging animals fully compensate for reduced forage availability in one area by transferring their attentions to other areas, when they may also be constrained by the need for water. It is probable that different species can compensate to differing degrees, and cattle may be better at this than sheep. Ultimately there are areas of most Australian pastoral landscapes which are either usually physiologically out of reach or totally rejected by free-ranging animals, so that grazing pressure in most paddocks ranges from very high near waterpoints and favoured vegetation types to very low elsewhere (Lange and Willcocks, 1980; Stafford Smith, 1988; Andrew, 1988). Herded animals may graze an area more uniformly, depending on the behaviour of and interactions between the herders.

Patterns of vegetation change in space and time

Given an understanding of the components of vegetation change, it is possible to demonstrate how complex spatial patterns can be generated from relatively simple functional responses applied to natural landscapes as a set of 'overlays'. This process illustrates the significance of the spatial dimension, both for management and for monitoring.

Firstly, consider the area of landscape which develops the patterns of water re-distribution shown in Plate 12.1. In conjunction with ground mapping, satellite remote sensing can be used to classify the landscape into major land units as shown in Plate 12.2A. With the aid of GIS technology, the locations of waterpoints and fencelines can be superimposed, and the distance from water of all points calculated (Plate 12.2B). The likely patterns of impact from cattle grazing (Plate 12.3A) can be predicted from these factors and the pattern of vegetation cover (Plate 12.3B), and a forecast can be made of the distribution of erosion and deposition (Plate 12.4).

From the patterns of Plate 12.4, we can map areas where changes in vegetation may occur. Depositional zones subject to grazing are at risk of shrub encroachment (Plate 12.4A), whilst other areas need monitoring and management for high erosion risk (Plate 12.4B). In fact, simple models can suggest much more. Depositional areas, with better resource continuity and reliability, show a trend in an almost Clementsian fashion towards perennial plants, and, if left 'undisturbed', probably towards a woody overstorey. Of course, these areas usually *are* 'disturbed' even in the ungrazed state, by fire and by intermittent extreme droughts which act against this trend. By comparison, source zones have poor water retention as well as unstable soils, and poor nutrient and seed supplies; they tend to carry ephemeral or short-lived plants, but also very hardy surviving perennials on residual islands of topsoil. Transfer zones, with longer periods of resource stability, fall between the two, and carry perhaps the widest range of plant strategies in response to the variety of opportunities, from short-lived plants to scattered long-lived perennials which establish in transient depositional areas.

If we now add the cattle grazing pattern, unpalatable plants will be favoured over palatable ones in areas of high impact; this effect will be more significant on perennial plants than short-lived ones, and there will be a feedback effect as sinks intensify, since these will become the more reliable sources of plant production. Increased grazing in the sinks may now select against grasses, suppress fire and enhance the development of woody plants to the extent that they dominate. Grazing may then be excluded because accessibility is limited and fodder plants are no longer present. Thus, from simple processes, it is possible to understand the development of a complex landscape pattern of palatability changes.

Landscape models such as these are currently being built from their constituent parts and compared with the changes recorded in archival satellite data, over large areas in central Australia. This technique can be used to test whether the models make sense, and how sensitive the observed changes are to each model component.

The processes which we used to build the complex spatial pattern can also be reversed to extract the effects of a particular driving force such as grazing (Fig. 12.3). Until recently, grazing patterns were difficult to measure because of the need for expensive and labour-intensive techniques such as dung counts, radio-tracking or aerial survey (Stafford Smith, 1984; Newsome et al., 1981). However, it has now been shown that, in large paddocks at least, grazing patterns can be inferred from changes in vegetation cover over time observed in satellite imagery (Pickup and Chewings, 1988b; Pickup, 1993). These patterns may be used as spatial 'filters' to separate natural and induced changes in vegetation characteristics.

The grazing gradient technique (Pickup, 1989; Pickup and Chewings, 1993) uses this approach and assumes that grazing intensity can be related to distance from water in a given vegetation type. Vegetation parameters which are affected by grazing will vary systematically with distance from water out to a certain point, beyond which they become constant. The constant element

(text continues on page 218)

Figure 12.3: Alternative modelling processes

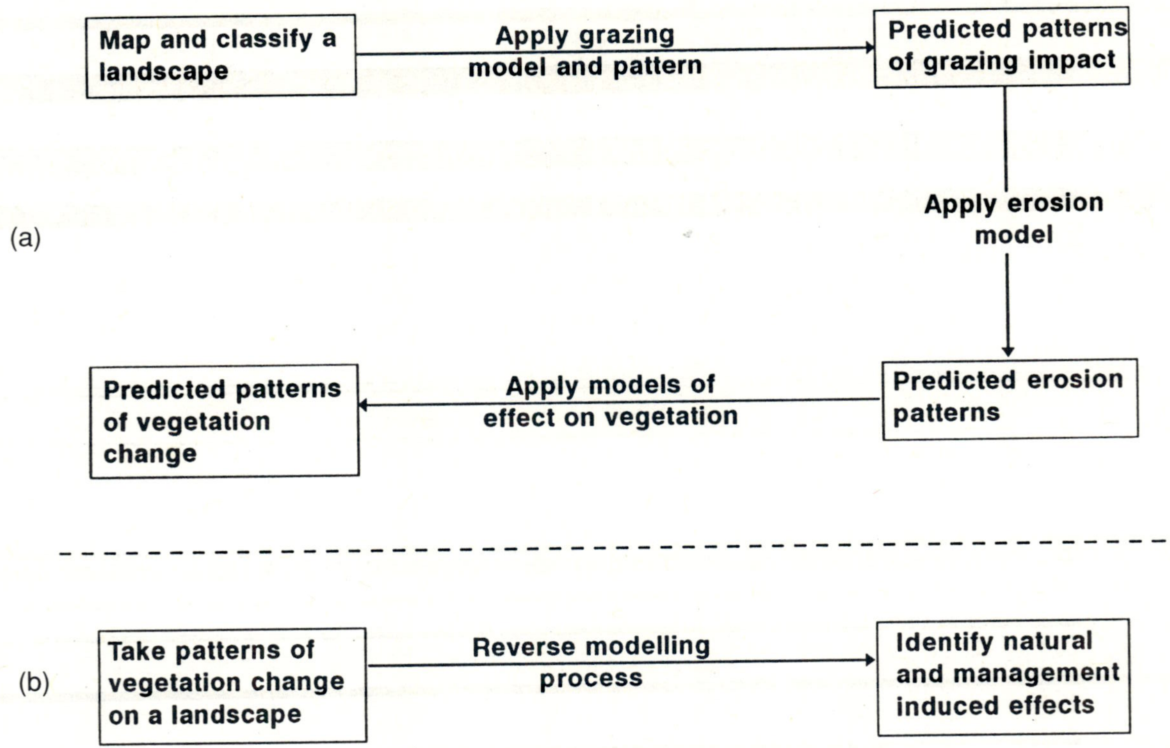

Notes: The modelling process followed through the sequence of Plates 12.1-12.4 to develop complex landscape patterns out of relatively simple driving forces, indicating how the procedure is then reversed to extract the natural and management imposed components of the forces from the observed complexity.

Plate 12.1: Modelled water re-distribution patterns in areas with significant topography

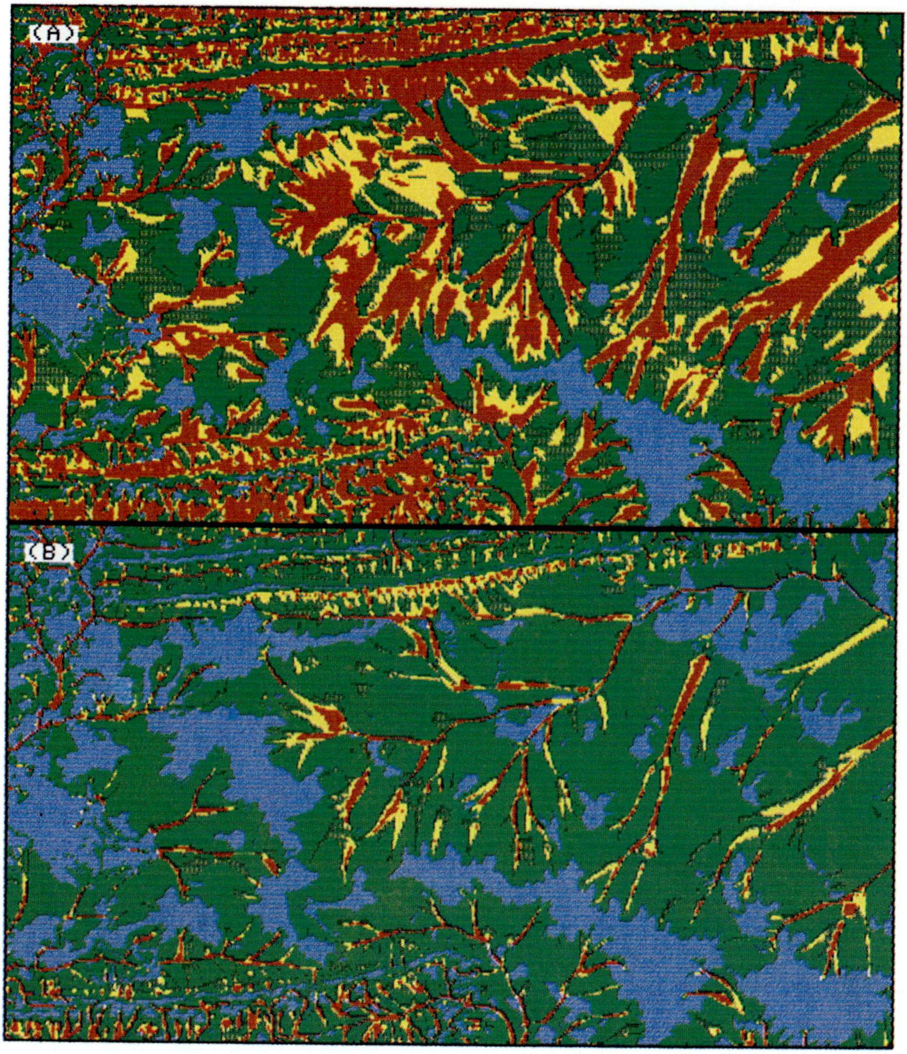

Key: greens - shedding run-off, yellow and red - receiving run-on, blues - no significant water re-distribution. See Notes on page 217.

Plate 12.2: The area of central Australia shown in Plate 12.1 classified using remote sensing and cadastral data

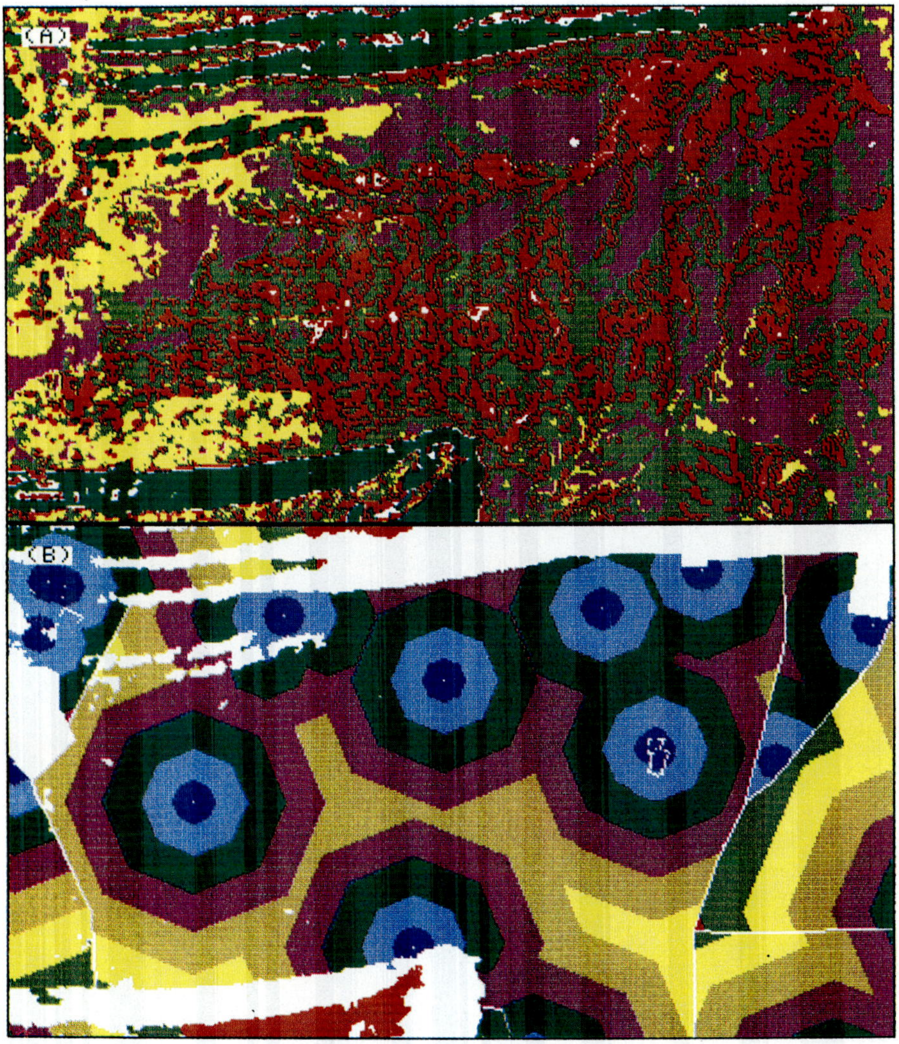

Key: (A) colours represent different vegetation types; (B) blues through green, purple, and yellows to red represent increasing distance from water, white is inaccessible to cattle. See Notes on page 217.

Plate 12.3: The predicted pattern of (A) grazing impact by cattle compared with (B) the observed pattern of vegetation cover

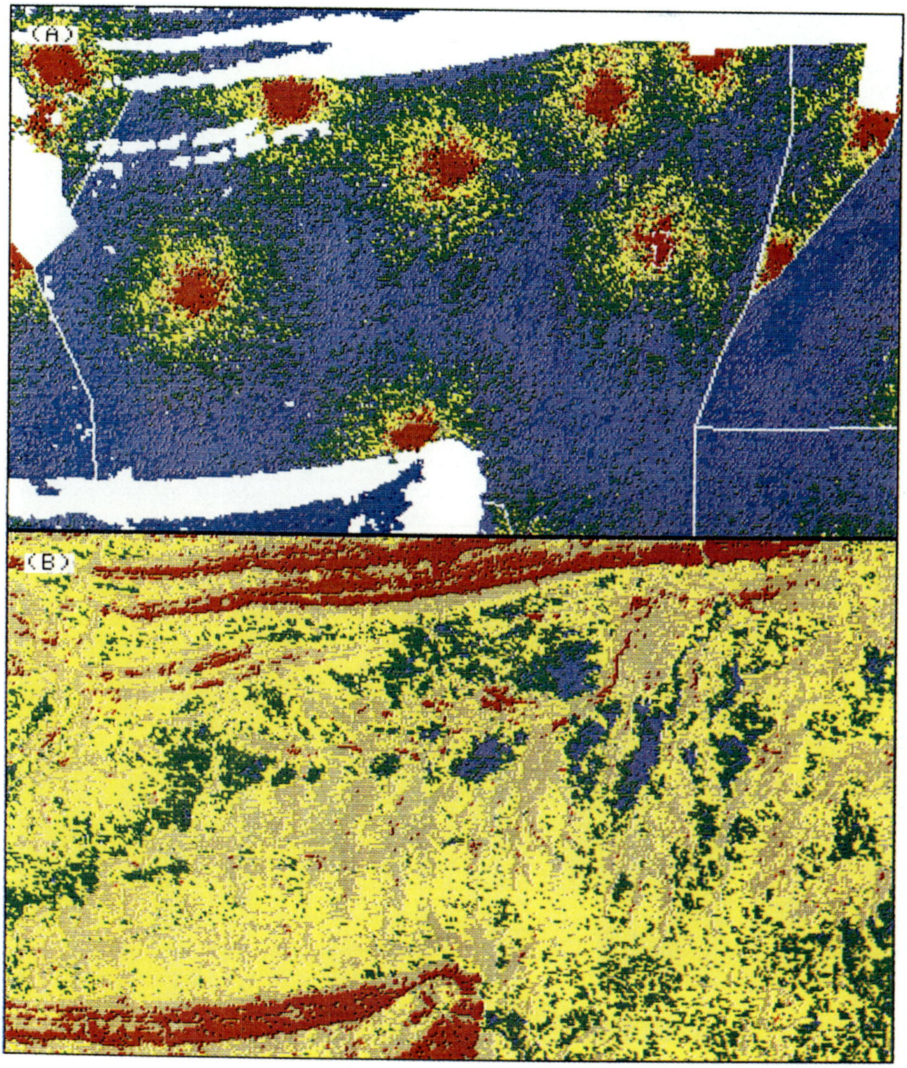

Key: (A) red greatest, through brown, yellow and green to blue, least, with white indicating no access by cattle; (B) red greatest through to blue least. See Notes on page 217.

Plate 12.4: Patches where changes important for monitoring and management may occur in the same area as previous plates

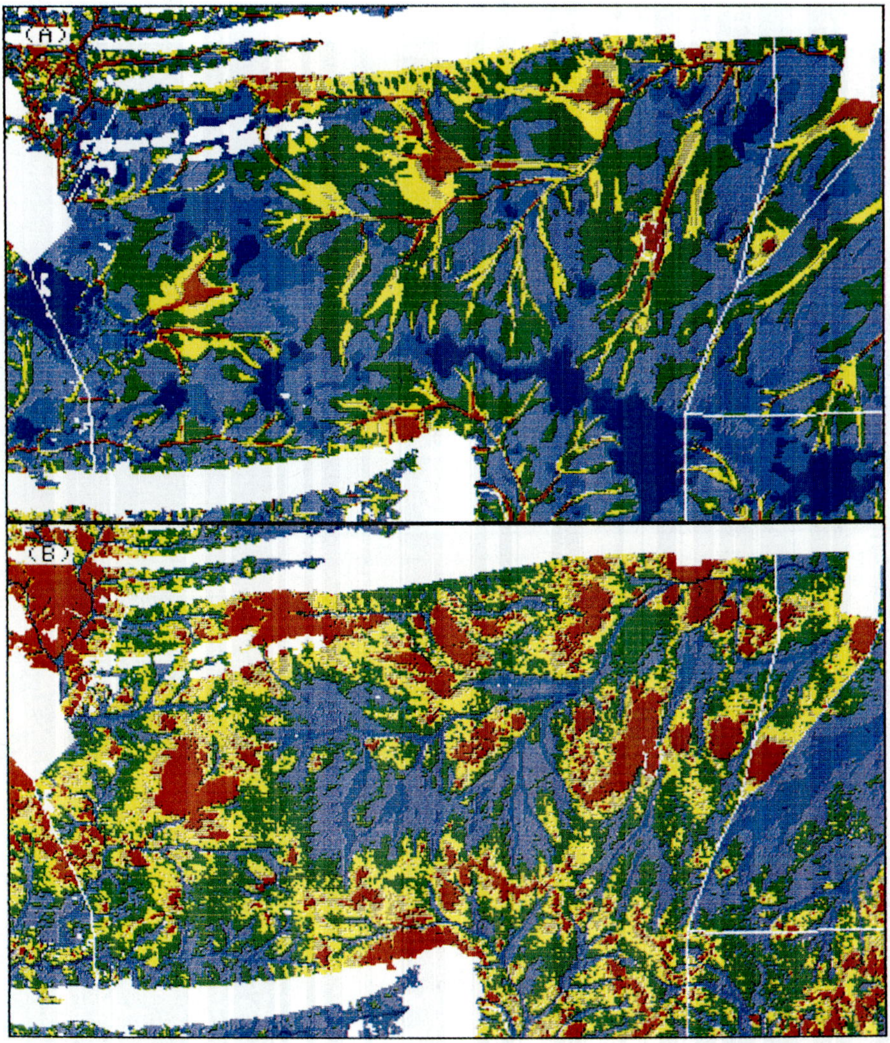

Key: Red denotes high risk through brown, yellow, green and light blue to dark blue indicating low risk. See Notes on page 217.

Notes to Plates

Plate 12.1

Map (A) is the pattern for a given rainfall in a dry period, (B) is the pattern for the same rainfall in a wet period when there is more vegetation. These patterns can change depending on the amount of vegetation present at the time of the rain; this mainly involves a shift up- and/or down-slope, as well as changes in the amplitude of the pattern especially in concave and convex elements of the landscape. Considerable research has been done on topographically-controlled processes of water and sediment re-distribution, and predictive methods are now reaching the operational stage. The principal restriction in the use of these techniques is the need for a high quality Digital Elevation Model, itself dependent on a comprehensive data set of altitudes and stream lines.

Plate 12.2

Map (A) shows the major landscape types as classified from remotely sensed data (green - mountain ranges, red - alluvial fans and depositional sinks, brown - mid-slopes with an aeolian mantle, purple - upper slopes with calcareous soils, yellow - as purple but with extensive erosion); note how spotty reality is compared with the idealised patterns shown on conventional maps. Map (B) uses waterpoint locations, fencelines and impassable hills (in white) to determine the effective distance from water of different areas.

Plate 12.3

Pattern on the central Australian landscape shown in Plates 12.1 and 12.2. The grazing pattern models relate to distance to water, vegetation type and cover with parameters derived from analysis of multitemporal satellite data.

Plate 12.4

Map (A) areas subject to shrub encroachment due to good water supply and significant grazing, and (B) areas at risk of serious erosion, due to a combination of grazing pressure and a natural propensity to erode. The patterns of erosion and shrub encroachment are an amalgam of the underlying landscape structure and the extra, spatially-variable pressure applied by cattle grazing. Loss of cover and reduced infiltration rates under heavy grazing and trampling pressure increase erosion in susceptible areas, producing an increase in the proportion of the landscape experiencing erosion. However, mean size of erosion cell and the variance in erosion-deposition sequences also increase, and deposition from eroding areas may occur at some distance from the areas under heavy grazing. The resulting pattern is not simple, even though it is generated by relatively simple relationships.

indicates the background or natural value for that parameter, while the spatially varying component indicates the grazing effect. Grazing gradient techniques are currently being used with vegetation pulses as recorded on satellite imagery to identify where grazing has an impact (cf. Plate 12.3). For example, systematic variations in the peak of the largest vegetation pulses suggest that, even in the best conditions, the ability of vegetation to respond to rainfall has been reduced by grazing in areas of high impact; the technique quantifies these areas. Similarly, trends in the variance of that peak with distance from water imply grazing-induced differences in the level of erosion cell development. Information on the spatial pattern of cover depletion after the rainfall pulse can also be used to infer change. If the rate of cover depletion varies systematically with distance from water, it can be used to derive the grazing distribution pattern. Given the known tendency of animals to disperse more when fodder is scarce (Arnold and Dudzinski, 1978), that pattern is an indicator of the amount of forage present. Examination of a number of production pulses may then be used to assess trends in forage quality over time.

Conceptual models of vegetation change

It is apparent from the above discussion of processes that, whilst it would be nice to be able to write straightforward equations which relate vegetation change to an earlier state and the values of the driving forces in the intervening time period, the responses are too complex for generalised models to be feasible, given our current understanding. The interactions are also too complex to expect managers (or anyone else) to conceptualise them in everyday decision-making. We shall return at the end of this section to considering which would be the best underlying concept or paradigm, given our inability to completely define process relationships. First, we briefly review the simplifications that have been, or are being, used to cope with the complexity of managing rangelands.

Clementsian paradigm

Range management in North American rangelands has rested on the Clementsian succession model for half a century. This model saw vegetation change as a simple linear progression from a newly disturbed, colonist-dominated vegetation composition through to a single climax composition, with movement backwards and forwards along this sequence without much hysteresis. This made for a straightforward management paradigm in rangelands, where increased grazing pressure was supposed to push the system towards the disturbed end of the continuum, and the release of that pressure would permit a balanced return to the climax vegetation which was presumed to be the ideal for a given site.

The majority of habitats do not fit the Clementsian model, and the model has particularly severe limitations when applied to episodically-driven arid and semi-arid rangelands. The continuum is not usually linear, there may be multiple 'climax' states which may not be very productive, different forms of 'disturbance' do not necessarily act in similar ways, and the removal of the disturbing force does not automatically allow a site to resume its orderly successional progression (Westoby et al., 1989). Perhaps most importantly as far as management is concerned, some sites never even approach a climax state because the so-called 'disturbances' (impulses from the driving forces, in our terminology) are so frequent that any directed succession can never progress to completion.

Although the model has been much maligned recently, and probably does not apply perfectly to any vegetation type, there are vegetation-climate combinations, such as the so-called 'climax-grasslands' of South Africa and the United States, in which it is a very useful simplification for management. The problem has arisen when the model has been taken as a paradigm from these areas and forced to fit other areas where its simplifying assumptions (not always clearly enunciated) were not appropriate; it is instructive to bear this problem in mind, to avoid the same problem with any new paradigm.

Multistate approaches

As mentioned in the introduction, a number of attempts have been made to address the problems of the Clementsian succession model in recent years. A more generalised approach is now emerging in which considerations of stability and resilience, soil-water behaviour, relative plant growth rates and the impact of grazing suggest that more than one stable condition may exist (Noy-Meir, 1975; Walker et al., 1981; Noy-Meir and Walker, 1986). Furthermore, the presence of different growth opportunities under complex weather sequences, the impact of vegetation change on soil properties, and the tendency for initial conditions to affect subsequent behaviour all mean that different growth forms can coexist in a stable fashion and that plant behaviour may follow different trajectories through time (Westoby, 1980).

The complexity of these ideas makes them difficult to use for practical land management but they can be captured and simplified for specific situations by being expressed as 'state and transition' models (Westoby et al., 1989). These models describe rangeland behaviour through catalogues of alternate states and transitions between states, each with a range of probabilities of occurrence and probabilities of reversal. Managers are then able to concentrate on timely intervention, in the sense that intervention at some times is pointless, at others critical. This contrasts with the Clementsian paradigm which implies that relief from disturbance at any time enables recovery. The state and transition model recognises that, for example, the removal of grazing may not cause any shift from a degraded condition, unless and until some combination of other conditions, such as fire and rain, allows the re-establishment of lost plants. In short, the model focuses on the 'event-driven' nature of many rangelands environments (Griffin and Friedel, 1985).

These new approaches are more holistic and realistic than earlier models when applied to arid and semi-arid areas with erratic rainfall. The holism arises from explicit consideration of linkages and feedbacks described earlier - between soil behaviour, plant behaviour, grazing and the impact of complex climatic sequences. Changes in, and modification of, the driving forces therefore become part of the process-response model, whereas in the classic range succession model they are treated as external. Holism adds realism to the new models but it also exposes their incompleteness. This incompleteness arises because the models are formulated to deal with points in a landscape or landscape averages, yet rangeland managers deal with whole landscapes often having great spatial variability. Although the state and transition approach can be 'stretched' to encompass spatial variation to some extent, it is not a part of the formulation to imply that a change in one place may be causally linked to a different change elsewhere. Unless the forms of spatial and temporal variability we have described earlier are explicitly built into models of vegetation dynamics, the ability of those models to explain, predict and provide frameworks for land management will be severely limited.

Towards a broader concept

There are two problems with state and transition-based approaches at present - the risk that the simplification for management becomes the paradigm, as happened with the Clementsian model, and the failure to deal explicitly with spatial patterns and flows where necessary.

To approach the problem of spatial pattern, consider the following ideas (developed from Stafford Smith, 1992a). It remains the case that a perfect process understanding (as well as large computers and some sophisticated stochastic models) would enable us to define the state of vegetation at a point or on a landscape in terms of its state at a previous time and the driving forces of climate, grazing, fire and resource redistribution in the intervening period. We cannot yet expect to do this except for a few simple systems or system components, so we must develop a simpler abstraction or model. In recent years various workers (Austin et al., 1981; Foran, 1986; Foran et al., 1986; Peco, 1989) have plotted multi-temporal assessments of site vegetation composition on to an ordination, where the axes are some function of the site descriptors. They have thus shown how the site 'moves around' in this 'n-dimensional species space' through time, with the interesting result that the trajectories of some sites are quite different in nature from others.

This seems to be a powerful way of visualising the behaviour of the vegetation composition of a site, which can be considerably extended. Effectively one could consider what positions a site could ever occupy if it is plotted in n-dimensional space, where the dimensions are different site descriptors (pragmatically these would usually be the presence/absence or percentage compositions of each plant species that ever occurs at the site, but they could be any site descriptor including factors such as soil type, diversity, biomass or even invertebrate species). Now imagine that the density of the plot at any given location is proportional to how frequently that point was occupied if the site had been monitored monthly over many thousands of years. Some sort of probability density function in n-dimensions results, as illustrated in Fig. 12.4, where the situation has been simplified down to 2- and 3-dimensions.

If all possible combinations of species were equally likely, the distribution would be uniform in the n-dimensions (Fig. 12.4a). The Clementsian succession model presumes that the distribution is in some way linear (not necessarily

Figure 12.4: Hypothetical examples of different probability distributions of a site in a 2- (left) and 3- (right) dimensional descriptive space (after Stafford Smith, 1992a)

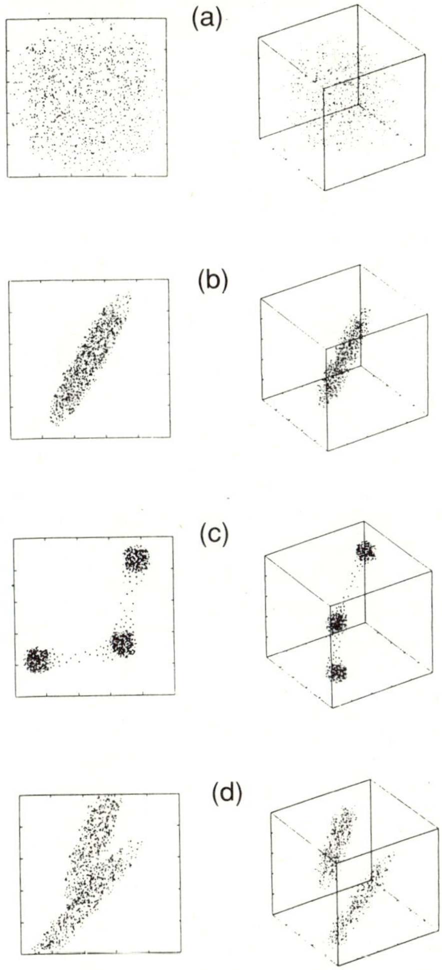

Notes: (a) an unlikely distribution where all combinations of descriptors are equally probable, (b) a site capable of Clementsian-type succession along a single path (by implication all types of 'disturbance' must act to move the site within this probability distribution, but no indication is given of whether assumptions such the lack of hysteresis in the rates of succession are satisfied), (c) a site capable of oscillating in a 'stable' fashion within 3 different 'domains of attraction', or, under some influence, transiting rapidly between domains, as occurs under the state-and-transition model (the pattern suggests that a direct transfer between two of the states cannot happen in this case), (d) a possible more complex pattern, with two alternative linear successional paths (note how they apparently intersect on the basis of 2 descriptors, but are clearly separate paths when a 3rd descriptor is included); these could be a result of changing erosional conditions at the point.

straight, but at least linked on what is topologically a single line) (Fig. 12.4b). For the state and transition model, there are a number of regions in the n-dimensions with a high probability density, with links of low probability density between some of them (Fig. 12.4c). The density of these linking points reflects how quickly the change between states occurs, and how likely it is to occur, thus highlighting the fact that the 'transition' of the state and transition model is a continuum from things which happen catastrophically fast on a management timescale (e.g. fire and drought), through slower events (e.g. the invasion of unpalatable perennial plants), to long-term vegetation change not necessarily very different from Clementsian succession. Importantly, it is possible to imagine all sorts of other possible distributions, such as a divaricating sausage, or two separate sausages implying two separate Clementsian-type streams with a state and transition type link between them (Fig. 12.4d).

So far we have not considered spatial changes in discussing Fig. 12.4; however, Fig. 12.4d could describe a point in an erosion zone and include the changes which result from increased erosion intensity across the landscape - that is, the two 'sausages' could represent the vegetation responses of the point when it happened to be in the source zone of a passing erosion cell, and the other when it was in the sink zone. Such behaviour could be quite common for points in transfer zones where they regularly switch between gaining and losing sediment (see Pickup, 1985). Thus Fig. 12.4 can encapsulate some of the spatial variability of the landscape. The amount that is included depends on the rate of landscape changes at a point. Thus whilst a point on a floodplain might behave like Fig. 12.4d where substantial soil changes are slow or even fast processes, a nearby point on stable ground above the floodplain might behave like Fig. 12.4a or b with resource changes being slow or static processes.

In fact, the concept of Fig. 12.4 can be extended further to the extent that it no longer represents the possible states of a single point, but a statistical description of an area of landscape, where that statistical description includes feedback relationships between landscape components. At this stage the scale of the sampling area becomes very important; the foregoing discussion presumes a point sample which, on a long enough time scale, encompasses both inherent vegetation variability and the variability resulting from changes in resource availability at the point. Expanding the concept to describe an area engulfs also the linkages between different components of that area.

An example may help to illustrate this. According to erosion cell theory, the frequency of points in a depositional state in the landscape should be related to the frequency of points in an erosional state. Eroding areas tend to have less vegetation cover than depositional ones, so there are effectively two frequency distributions of cover. When these are combined, they spread across a wider range of cover values, so the variance of the combined distribution is higher than that of its individual components (Fig. 12.5a). Landscape variability can also change through time. Fig. 12.5a represents conditions during a very wet time; almost all parts of the landscape develop substantial cover, blurring the distinction between eroding and depositional areas. After a period of natural drying and grazing, however, the difference becomes much clearer (Fig. 12.5b), showing how a spatial factor such as erosional state can generate patterns akin to those of Fig.12.4d, and emphasising the significance of spatial scale in vegetation change.

What is the value of this view of vegetation change? The importance for those who produce models for management is that the Clementsian and the state and transition models can be readily visualised as simplifications of a much larger set of possible types of distributions. Users should, therefore, be more inclined to use these simplifications only in situations where they apply, rather than forcing all data to fit one or other model. If a particular site really behaved like Fig. 12.4d, either of the simpler models would fail since not only do the transitions between the 'sausages' need the opportunistic management approach associated with state and transition models, but changes within the 'sausages' are substantial enough to need managing too, whether these changes represent behaviour at a point over time or the spatial linkages of an entire landscape.

The problem with the situation shown in Fig. 12.4 is that rates of change are not necessarily highlighted, although the density of the diagram partly indicates this. Ultimately we need to focus on what driving forces cause what shifts through the n-dimensional distribution, and how quickly: we have to recall that a manager's threshold may be a policy-maker's continuum, so that an understanding of rates of change in relation to management time-frames is crucial. The state and transition model will always be a useful simplification in appropriate systems, but users need to be continually reminded of the necessary spatial context with its vital resource fluxes. Hence, we believe that with further development the approach shown in Fig. 12.4 may be a useful way to consider the spatial and temporal problems of vegetation change in the rangelands.

What is the relevance to African systems?

Australian rangelands differ from African rangelands at ecological, economic, social and cultural levels. The present authors do not claim great knowledge of the African situation but we believe that Australian work *is* relevant, given some caveats, for the following reasons.

Ecological system

We presume that, given similar soils and driving forces, the ecological principles applying in different parts of the world should be the same. Problems arise when the principles are not truly universal and hide local assumptions. Stafford Smith and Morton (1990) delineate some physical aspects of the Australian system which may be unusual elsewhere in the world; these include the exceptionally variable and unreliable climate, the extensiveness of areas with very poor soils and the importance of sorting on this ancient and flat continent. All of these features occur in parts of Africa, but perhaps not to the same combinatorial degree or areal extent. However, these features all emphasise the importance of variability as discussed in this

Figure 12.5: Distribution of pixels with different vegetation cover index values for two of the land units shown in Plate 12.2 (A)

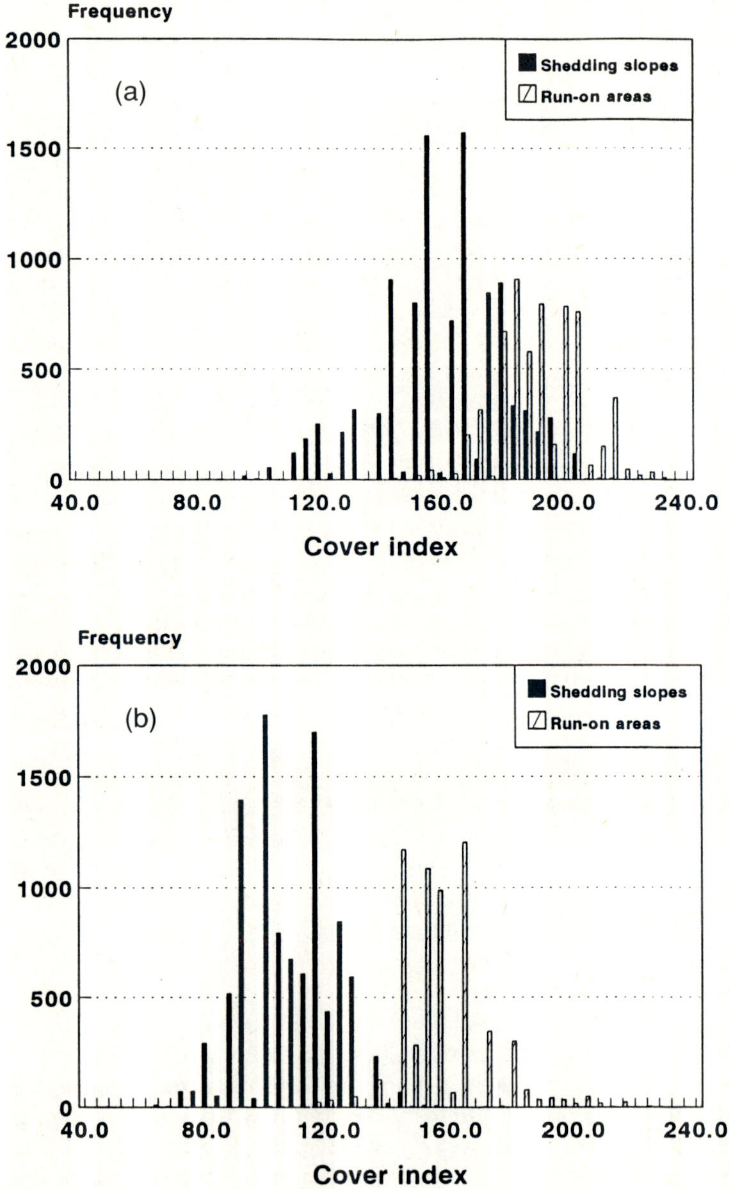

Notes: (a) a wet period, (b) a dry period, illustrating how the behaviour of different parts of the landscape diverges (the frequency distributions overlap less) during a drying phase. The cover index is derived from Landsat MSS data and varies between 0 and 254 with 254 representing 100% cover.

chapter, rather than reducing the attention paid to other factors. Studies such as Scoones (1989b) draw attention to different components of the landscape which parallel those described earlier for Australia very closely.

An important ecological difference between most African and Australian vegetation may be the absence of significant numbers of browsers for the last 10-30,000 years in Australia (Flannery, 1990). This may have resulted in a relaxation of many selective pressures on plants, now re-imposed by introduced horses, cattle and sheep. The fact that a substantial suite of productive but palatable plants seem very susceptible to overgrazing may colour Australian approaches to grazing management, where there is still the option of retaining these components of the vegetation. By comparison, African systems may have come to a new 'equilibrium' after losing the equivalent suite long ago, either from the pressure of the natural herds (possibly enhanced by a slightly more reliable climate or migratory habits) or from domestic herding (cf. Walker et al., 1987; MacDonald, 1985). Some African vegetation also seems to be more substantially armoured with thorns and spines (cf. references in Stafford Smith and Morton, 1990), although it is still unclear whether this results from underlying differences in resource availability or historical selection pressures (cf. Milton, 1991).

Economics and the socio-cultural system

There are obvious differences between the capitalist pastoral enterprise system in Australia and the subsistence economy in Africa, although there may be fewer differences between Australian and African ranch enterprises. In general, though, Australian pastoralists replace high labour inputs with comparatively enormous land areas and high technology such as helicopter mustering. Some management options which are uneconomic in Australia are therefore quite feasible in Africa, including control of feral animals and use of 'undesirable' woody plants. African pastoralists also have the option of controlling the evenness of grazing far more efficiently than Australians, although this does not change the importance of differential usage across the landscape as a principle.

On both continents, managers are greatly affected by climatic variability; until recently Australian pastoralists probably accepted overgrazing and eventual great stock losses at least as much as African pastoralists. Today, drought management in Australia is slowly improving, and this results in a greater degree of 'hi-tech transhumance' than is commonly realised, either by selling animals to other regions, or by moving animals to be fed on properties elsewhere which may even be owned by the same company. Recent work from Africa suggests that subsistence herders are more tightly involved in the economy than has been hitherto assumed (Behnke, 1987); equally, Australian pastoralists have many enterprise goals which are not directly related to economic returns (Stafford Smith 1992b). The importance of clearly defining what the real goals of management are is evidently equally important on either continent, and has probably been given less than the desirable level of attention on both.

Conclusions

Variability in space and time is a particularly critical factor in the rangelands, given the scale at which they must be managed. It has lead to serious problems with land degradation and with policy paralysis in the past. This is hardly surprising in retrospect, but today's challenge is to try to change this situation, both on the ground and in policy. The implications for managers are dominated by the need to cope with risk. There seem to be a limited number of different strategies available (Stafford Smith, 1992b), paralleling the 'conservative' and 'opportunistic' approaches described by Sandford (1983a) in Africa. However, there are many tactical ways of implementing these strategies, driven more by economics than ecology, and a better understanding of these is still needed. The policy side must deal with the (potentially healthy) tension between the rights of individual managers who have access to leasehold land, and their responsibilities as monitored by government in its role as the protector of the broader interests of society. The problems of monitoring large management units in a variable climate have been exacerbated for government by poor models of landscape function. There is now a clear role for some hierarchical system of land monitoring, coupling remotely-sensed data at a variety of scales with intensive ground-based monitoring (e.g. Hacker et al., 1990) which can never by itself provide full coverage (Pickup, 1989). Finally, the social, cultural and bureaucratic inertia acting against efficient and effective decision-making should never be underestimated, in Australia as elsewhere. Despite this, Australians are beginning to capitalise on the developments described in this chapter to improve the management of their rangelands.

References

à Tchie, N. and Gakahu, G.E. 1989 Responses of important woody species of Kenya's rangelands to a prescribed burning. *African Journal of Ecology* 27: 119-128.

Abel, N.O.J. 1993 *What's in a Number? The Carrying Capacity Controversy on the Communal Rangelands of Southern Africa.* Ph.D. thesis, University of East Anglia, Norwich.

Abel, N.O.J. and Blaikie, P. 1989 Land degradation, stocking rates and conservation policies for the communal rangelands of Botswana and Zimbabwe. *Land Degradation and Rehabilitation* 1: 101-123. Also published as *Pastoral Development Network* No. 29a. London: Overseas Development Institute.

Abel, N.O.J., Flint, M.E.S., Hunter, N.D., Chandler, D., and Maka, G., 1987 *Cattle-keeping, Ecological Change and Communal Management in Ngwaketse.* Addis Ababa: International Livestock Centre for Africa, and Gaborone: Ministry of Agriculture.

Abel, N.O.J., and Stocking, M.A. 1987 A rapid method for assessing rates of soil erosion from rangeland: An example from Botswana. *Journal of Range Management* 405: 460-466.

Acocks, J.P. 1953 Veld types of South Africa. *Botanical Survey of South Africa* Mem. 28.

Agritex n.d. Livestock carrying capacity recommendations by natural region. Mimeo. Harare: Agritex, MLARR.

AGROTEC/CRG/SEDES Associates. 1974a Southern Rangelands Livestock Development Project. Part II. *Studies and Surveys Volume 1: Range Ecology.* Rome: Commissioned by the Livestock and Meat Board of the Ethiopian Government.

AGROTEC/CRG/SEDES Associates. 1974b Southern Rangelands Livestock Development Project. Part II. *Studies and Surveys Volume 4: Surface Hydrology and Geohydrology.* Rome: Commissioned by the Livestock and Meat Board of the Ethiopian Government.

AGROTEC/CRG/SEDES Associates. 1974c Southern Rangelands Livestock Development Project. Part II. *Studies and Surveys Volume 4: Surface Hydrology and Geohydrology Appendix - Statistics.* Rome: Commissioned by the Livestock and Meat Board of the Ethiopian Government.

Ahlchrona, E. 1988 *The Impact of Climate and Man on Land Transformation in Central Sudan: Applications of Remote Sensing.* Avhandlingar 103: Meddelanden fran Lunds Universitets Geografiska Institutioner.

Andrew, M.H. 1986 Use of fire for spelling monsoon tallgrass pasture grazed by cattle. *Tropical Grasslands* 20: 69-77.

Andrew, M.H. 1988 Grazing impact in relation to livestock watering points. *Trends in Ecology & Evolution* 3: 336-339.

Anon. 1976 Soil loss estimator for southern Africa. *Bulletin No. 7 of the Natal Agricultural Research Institute.* Pietermaritzburg: Department of Agricultural Technical Services.

Antenneh, A. 1984 Trends in sub-Saharan Africa's livestock industries. *ILCA Bulletin* 18: 7-15.

APRU (Animal Production Research Unit) 1975-1988 *Livestock and Range Research in Botswana.* Annual Reports of the Animal Production and Range Research Unit. Gaborone: Ministry of Agriculture.

APRU (Animal Production Research Unit) 1980b *Beef Production and Range Management in Botswana.* Animal Production Research Unit. Gaborone: Ministry of Agriculture.

Archer, S. 1989 Have southern Texas savannas been converted to woodlands in recent history? *American Naturalist* 134: 545-561.

Arnold, G.W. and Dudzinski, M.L. 1978 *Ethology of Free-Ranging Domestic Animals.* Amsterdam: Elsevier.

Arntzen, J.W. and Veenendaal, E.M. 1986 *A Profile of Environment and Development in Botswana.* Amsterdam: Institute for Environmental Studies, and Gaborone: National Institute of Development Research and Documentation.

Assefa Eshete, Billé, J.C. and Corra M. 1986 *Ecological Map of South-Western Sidamo.* JEPSS Joint Ethiopian Pastoral Systems Study Research Report No. 19. Addis Ababa: ILCA (International Livestock Centre for Africa).

Austin, M. P., Williams, O.B. and Belbin, L. 1981 Grassland dynamics under grazing in an Australian Mediterranean type climate. *Vegetatio* 46: 201-211.

Barnes, D.L. 1965 A stocking rate trial in the Rhodesian high altitude sand veld. *Rhodesian Journal of Agricultural Research* 3: 101-107.

Bartels, G.B., Norton, B.E. and Perrier, G.K. 1990 The applicability of the carrying capacity concept to Africa: comment on PDN paper 29b by de Leeuw and Tothill. *Pastoral Development Network* Paper No. 30e. London: Overseas Development Institute.

Bartolome, J.W. 1984 Impacts of grazing intensity and grazing systems on vegetation composition and production. In *Developing Strategies for Rangeland Management.* A report prepared by the Committee on Developing Strategies for Rangeland Management, National Research Council. Boulder, Colorado: Westview Press.

Behnke, R.H. 1985a Measuring the benefits of subsistence versus commercial livestock production in Africa. *Agricultural Systems* 162: 109-135.

Behnke, R.H. 1985b *The Dynamics of Open-Range Management and Property Rights in Pastoral Africa.* mimeo, London: Overseas Development Institute.

Behnke, R.H. 1987 Cattle accumulation and the commercialisation of the traditional livestock industry in Botswana. *Agricultural Systems* 24: 1-29.

Behnke, R.H. and Kerven, C. 1984 *Herd Management Strategies among Agro-Pastoralists in the Bay Region, Somalia* Report for the Bay Region Socio-Economic Baseline Study, University of Wyoming and USAID (United States Agency for International Development)/Somalia.

Behnke, R.H. and Scoones, I. 1991 Rethinking range ecology: Implications for rangeland management in Africa. London: Commonwealth Secretariat, Overseas Development Institute and International Institute for Environment and Development.

Bell, R.H.V. 1982 The effect of soil nutrient availability on community structure in African ecosystems. In Huntley, B.J. and Walker, B.H. (eds.).

Bell, R.H.V. 1984 Notes on elephant-woodland interactions. In Cumming, D.H.M. and Jackson, P. (eds.) *The Status and Conservation of Africa's Elephants and Rhinos*, pp. 98-103 Gland: IUCN (International Union for the Conservation of Nature and Natural Resources).

Bell, R.H.V. 1985 Carrying capacity and offtake quotas. In Bell, R.H.V. and McShane Caluzi, E. (eds.) *Conservation and Wildlife Management in Africa.* Washington: US Peace Corps.

Belsky, A.J. 1989 Landscape patterns in a semi-arid ecosystem in East Africa. *Journal of Arid Environments* 17: 265-270.

Belsky, A.J. 1986 Does herbivory benefit plants? A review of the evidence. *The American Naturalist* 127: 871-892.

Billé, J.C. 1982 *The measurement of conditions and trends in the southern rangelands of Ethiopia.* JEPSS Joint Ethiopian Pastoral Systems Study Research Proposal No. 2. Addis Ababa: ILCA (International Livestock Centre for Africa).

Billé, J.C. 1985 Some aspects of bush encroachment in the African rangelands. In Tothill J.C. and Mott J.J. (eds.) *Ecology and management of the world's savannas.* Canberra: Australian Academy of Science.

Billé, J.C. n.d. *The measurement of changes in tropical rangeland ecosystems.* Addis Ababa, Ethiopia. ILCA (International Livestock Centre for Africa).

Billé, J.C. and Assefa Eshete 1983a *The ecology of Sarite and Dembala Wacha cooperative ranches.* JEPSS Joint Ethiopian Pastoral Systems Study Research Report No. 3. Addis Ababa: ILCA (International Livestock Centre for Africa).

Billé, J.C. and Assefa Eshete 1983b *Rangeland management and range condition: A study in the Medecho and Did Hara areas of the effects of rangeland utilization.* JEPSS Joint Ethiopian Pastoral Systems Study Research Report No. 7. Addis Ababa: ILCA (International Livestock Centre for Africa).

Billé, J.C. and Assefa Eshete 1984 *The marginal lands of western Sidamo.* JEPSS joint Ethiopian Pastoral Systems Study Research Report No. 14. Addis Ababa: ILCA (International Livestock Centre for Africa).

Billé, J.C. and Corra M. 1986a *Nitrogen and phosphorus in the ecosystems of Sidamo.* JEPSS Joint Ethiopian Pastoral Systems Study draft research memo No. 25. Addis Ababa: ILCA (International Livestock Centre for Africa).

Billé, J.C. and Corra M. 1986b *Influence of Acacia Trees.* JEPPS Joint Ethiopian Pastoral Systems Study draft research memo No. 26. Addis Ababa: ILCA (International Livestock Centre for Africa).

Billé, J.C., Assefa Eshete and Corra M. 1983 *Ecology and ecosystems of the Borana deep wells area.* JEPSS Joint Ethiopian Pastoral Systems Study Research Report No. 13. Addis Ababa: ILCA (International Livestock Centre for Africa).

Biot, Y. 1988 *Forecasting Productivity Losses Caused by Sheet and Rill Erosion in Semi-Arid Rangeland. A Case Study from the Communal Areas of Botswana.* Ph.D. thesis, University of East Anglia, Norwich.

Biot, Y. 1990a 'THEPROM - An Erosion Productivity Model'. In Boardman, J., Foster, I.D.L. and Dearing, J.A. (eds.) *Soil Erosion on Agricultural Land.* Chichester, England. John Wiley & Sons.

Biot, Y. 1990b Can livestock production be sustained in the Hardveld of Botswana? *Pedologie* XL-3: 243-255.

Biot, Y. 1991 How long can livestock production be sustained in the Hardveld of Botswana? Forthcoming in *Pedologie.*

Biot, Y., Sessay, M. and Stocking, M. 1989 Assessing the sustainability of agricultural land in Botswana and Sierra Leone. *Land Degradation and Rehabilitation* 1(4): 263-278.

Blackburn, W.H. 1984 Impacts of grazing intensity and specialized grazing systems on watershed characteristics and responses. In the Report of the Committee on Developing Strategies for Rangeland Management. *Developing Strategies for Rangeland Management.* pp. 927-984. National Research Council/National Academy of Sciences. Boulder, Colorado: Westview Press.

Blackburn, W.H., Knight, R.W., Wood, M.K. and Merrill, L.B. 1980 Watershed parameters as influenced by grazing. *American Society of Civil Engineers: Symposium on watershed management in the 1980s* 1: 552-572.

Blair Rains, A. and Kassam, A.H. 1979 *Land Resources and Animal Production* Consultants' working paper No. 8 FAO/UNFPA Project INT/75/P13 AGLS. Rome: FAO (Food and Agriculture Organisation of the United Nations).

Blair Rains, A. 1983 A review of rangeland inventory and evaluation techniques in anglophane [sic] Africa. In Siderius (ed.), pp. 27-35.

Blench, R.M., Bourn, D. and Wint, D. 1985 *Integrated air-ground surveys in the Nigerian sub-humid zone.* Report to the Sub-humid Zone Programme. Kaduna, Nigeria: ILCA (International Livestock Centre for Africa).

Booysen, P de V. 1981 Intensification of grassland production. In Tainton, N.M. (ed.) *Veld and Pasture Management in South Africa.* Pietermaritzburg: Shuter and Shooter.

Bosch O.J.H and Van Wyk J.J.P. 1970 The influence of bushveld trees on the productivity of *Panicum maximum*: A preliminary report. *Proceedings of the Grassland Society of South Africa* 5: 69-74.

Botkin D.B, Mellilo J.M.and Wu L.S.Y. 1981 How ecosystem processes are linked to large mammal population dynamics. In Fowler, C.W. and Smith T.D. (eds.) *Dynamics of Large Mammal Populations.* New York: John Wiley & Sons.

Bouba-Kaou, A. 1990 Mesure de la charge pastorale et du comportement du bétail à la pâture sur quelques parcours du Gourma. Memoire de fin d'études de l'Institut Polytechnique Rural de Katibougou. Bamako, Mali: CIPEA (International Livestock Centre for Africa).

Boudet, G. 1978 Manuel sur les pâturages tropicaux et les cultures fourragères. *Série Manuels et Précis d'élevage* No. 4. IEMVT (Institut d'Élevage Médecine Vétérinaire Pays Tropicaux). Paris: Ministère de la Coopération.

Boudet, G., Cortin, A. and Macher, H. 1971 *Esquisse Pastoral et Esquisse de Transhumance de la Région du Gourma, Mali.* DIWI/IEMVT (Institut Élevage Médecine Vétérinaire Pays Tropicaux), Essen and Maisons Alfort.

Bourn, D. and Wint, F. 1985a *Wet season distribution and abundance of livestock populations and human habitation in the Office du Niger Region of Mali.* Jersey: RIM Resource Inventory and Management Limited).

Bourn, D. and Wint, F. 1985b *Wet season distribution and abundance of livestock population and human habitation in Gourma Region of Mali.* Jersey: RIM (Resource Inventory and Management Limited).

Bourn, D. and Wint, F. 1986 Pastoral condition in Central Niger following the 1983/84 drought. Addis Ababa: ILCA (International Livestock Centre for Africa).

Braithwaite, R.W. 1991 Aboriginal fire regimes of monsoonal Australia in the 19th century. *Search* 22: 247-249.

Breman, H. and de Ridder, N. (eds.) 1991 Manuel sur les pâturages des pays sahéliens. Paris: *Karthala.*

Breman, H. and de Wit, C.T. 1983 Rangeland productivity and exploitation in the Sahel. *Science* 221: 1341-1347.

Breman, H., Ketelaars, J.J.M.H. and Traoré, N. 1990 Un remède contre le manque de terre? Bilan des éléments nutritifs, production primaire et élevage au Sahel. *Sécheresse* 2: 109-117.

Breman, H. and Traoré, N. (eds.) 1986a *Analyse des Conditions de L'élevage et Propositions de Politiques et de Programmes.* République du Niger, SAHEL D86284. Paris: OCDE (Organisation Coopération Développement Économiques)/CILSS (Comité permanent inter-Etats du latte contre Sécheresse Sahel)/Club du Sahel.

Breman, H. and Traoré, N. (eds.) 1986b *Analyse des Conditions de L'élevage et Propositions de Politiques et de Programmes. Burkina Faso.* SAHEL D86300. Paris: OCDE (Organisation Coopération Développement Économiques)/CILSS (Comité permanent inter-Etats du latte contre Sécheresse Sahel)/Club du Sahel.

Breman, H. and Traoré, N. (eds.) 1987 *Analyse des Conditions de L'élevage et Propositions de Politiques et de Programmes. Mali.* SAHEL D87302. Paris: OCDE (Organisation Coopération Développement Économiques)/CILSS (Comité permanent inter-Etats du latte contre Sécheresse Sahel)/Club du Sahel.

Breman, H., van Keulen, H. and Ketelaars, J.J.M.H. 1983 Land evaluation for semi-arid rangeland: a critical review of concepts. In Siderius (ed.), 229-244.

Brookes, E.H. and Hurwitz, N. 1957 *The Native Reserves of Natal.* Natal Regional Survey, Vol 7. Cape Town: Oxford University Press.

Buckley, R. 1985 Development in the central Australian arid zone: current assessment. *Journal of Arid Environment* 8(3): 173-189.

Burke, I C, Schimel, D.S., Yonker, C.M., Parton, W.J., Joyce, L.A. and Lauenroth, W.K. 1990 Regional modeling of grassland biogeochemistry using GIS. *Landscape Ecology* 4: 45-54.

Butterworth, M. and de Ridder, N. 1983 Report on Working Group D: Concepts and methods determining carrying capacity. In Siderius (ed.), pp. 325-326.

Campbell, A.C. and Child, G. 1971 The impact of man on the environment in Botswana. *Botswana Notes and Records* 3: 91-110.

Carew, C.W. 1976 Stocking rate as a factor determining the profitability of beef production. *Rhodesian Agricultural Journal* 73: 111-115.

Carl Bro International 1982 *An Evaluation of Livestock Management and Production in Botswana with Special Reference to the Communal Areas,* Vols. I, II and III. Evaluation Unit, Gaborone: Ministry of Agriculture and the Commission of the European Communities European Development Fund.

Caswell, H. 1982. Life history theory and the equilibrium status of populations. *American Naturalist* 120: 317-339.

Caughley, G. 1976 Wildlife management and the dynamics of ungulate populations. pp. 183-246. In Coaker, T.H. (ed.) *Applied Biology* Vol. 1. London: Academic Press.

Caughley, G. 1977 *The Analysis of Vertebrate Populations.* London: John Wiley & Sons.

Caughley, G. 1979 What is this thing called carrying capacity? In Boyce, M.S. and Hayden-Wing, L.D (eds.) *North American Elk: ecology, behaviour and management*, pp.2-8. Laramie, Wyoming: University of Wyoming Press.

Caughley, G. and Lawton, J. 1981 Plant herbivore systems. In May, R. (ed.) pp. 132-166. *Theoretical ecology*. Oxford: Blackwell Scientific.

Caughley, G., Shepherd, N. and Short, J. (eds.) 1987 *Kangaroos: Their Ecology and Management in the Sheep Rangelands of Australia*. London: Cambridge University Press.

Charley, J.L. and West, N.E. 1975 Plant induced soil chemical patterns in some shrub-dominated semi-desert ecosystems of Utah. *Journal of Ecology* 63: 945-963.

Christian, C.S. and Stewart, G.A. 1968 Methodology of integrated surveys. pp. 233-280, in *Aerial Surveys and Integrated Studies: Proceedings of the Toulouse Conference*. Paris: UNESCO.

Cincotta, R.P., Gay, C.W. and Perrier, G.K. 1991 *New Concepts in International Rangeland Development: Theories and Applications*. Proceedings of the 1991 International Rangeland Development Symposium, Society for Range Management. Logan, Utah: Department of Range Science, Utah State University.

Cissé, S. and Gosseye, P.A. (eds.) 1990 Compétition pour des ressources limitées: Le cas de la cinquème region du Mali. Rapport 1: *Resources Naturelles et Population*. Wageningen: CABO (Centre for Agrobiological Research) and Mopti, Mali: ESPR.

Cissé, A.M. 1986 *Dynamique de la Strate Herbacée des Pâturages de la Zone Sud-Sahélienne*. Thèse Ph.D, Université Agronomique, Wageningen.

Clark, C. 1976 *Mathematical Bioeconomics: The Optimal Management of Renewable Resources*. New York: John Wiley & Sons.

Clements, F.E. 1916 *Plant succession: an analysis of the development of vegetation*. Washington: Carnegie Institute Pub. 242: 1-512.

Cobbing, J. 1976 The Ndebele under the Khumalos, 1820-1896. Ph.D. thesis, University of Lancaster, Lancaster.

Coe, M.J, Cumming, D.M. and Phillipson, J. 1976 Biomass and production of large African herbivores in relation to rainfall and primary production. *Oecologia* 22: 341-354.

Colinvaux, P. 1986 *Ecology*. New York: John Wiley & Sons.

Cook, C.W. and Stoddart, L.A. 1953 The quandary of utilization and preference. *Journal of Range Management* 6: 329-335.

Cook, H.J. 1983 The struggle against environmental degradation-Botswana's experience. *Desertification Control Bulletin* 8: 9-15.

Cooke, R.U. and Reeves, R.W. 1976 *Arroyos and Environmental Change in the American South-West*. London: Oxford University Press.

Cooper, S.M. and Owen-Smith, N. 1985 Condensed tannins deter feeding by browsing ungulates in a South African savanna. *Oecologia* 67: 142-146.

Coppock, D.L. 1985 *Feeding Ecology, Nutrition, and Energetics of Livestock in a Nomadic Pastoral Ecosystem*. Ph.D. thesis, Colorado State University, Fort Collins.

Coppock, D.L. 1990 Bush control: Opportunity for resource management and pastoral income generation? Addis Ababa: *ILCA Newsletter* 93: 10-11.

Coppock D.L. In press *The Borana Plateau of Southern Ethiopia: Synthesis of Pastoral Research, Development and Change*. Systems Study No. 5. Addis Ababa: ILCA (International Livestock Centre for Africa).

Coppock, D.L., Ellis, J.E. and Swift, D.M. 1986 Livestock feeding ecology and resource utilization in a nomadic pastoral ecosystem. *Journal of Applied Ecology* 232: 573-584.

Coppock D.L., Ellis, J.E and Swift, D. M. 1988 Seasonal patterns of activity, travel and water intake of livestock in South Turkana, Kenya. *Journal of Arid Environments* 14: 319-331.

Coppock D.L. and Mulugeta M. 1985 Summary of Borana encampment survey conducted 16 November to 15 December, 1985. Internal memorandum. Addis Ababa: ILCA (International Livestock Centre for Africa).

Coppock D.L., Swift, D.M. and Ellis, J.E. 1986 Seasonal nutritional characteristics of livestock diets in a nomadic pastoral ecosystem. *Journal of Applied Ecology* 23: 585-595.

Cornet, A. 1984 Utilisation de modèles simples de bilan hydrique et de production de biomasse pour determiner les potentialités de production de parcours en zone Sahelienne Sénégalais. In Siderius (ed.), pp. 207-228.

Cossins, N.J. 1985 The productivity and potential of pastoral systems. *ILCA Bulletin* 21: 10-15.

Cossins N.J. and Upton M. 1987 The Borana pastoral system of southern Ethiopia. *Agricultural Systems* 25: 199-218.

Cossins N.J. and Upton M. 1988a The impact of climatic variation on the Borana pastoral system. *Agricultural Systems* 27: 117-135.

Cossins N.J. and Upton M. 1988b Options for improvement of the Borana pastoral system. *Agricultural Systems* 27: 251-278.

Cotton, J.S. 1906 Range management. pp. 225-238. In *Yearbook of the Department of Agriculture 1906*. Washington, DC: US Department of Agriculture.

Coughenour, M.B. 1991 Spatial modeling and landscape characterization of an African pastoral ecosystem: A prototype model and its potential use for monitoring drought. In McDonald, J.M. (ed.) *Proceedings of the International Symposium of Ecological Indicators*. New York: Elsevier Applied Science.

Coughenour, M.B., Coppock, D.L., Ellis, J.E. and Rowland, M. 1990 Herbaceous Forage Variability in an Arid Pastoral Region of Kenya: Importance of Topographic and Rainfall Gradients. *Journal of Arid Environments* 19(2): 147-159.

Coughenour, M.B., Ellis, J.E., Swift, D.M., Coppock, D.L., Galvin, K., McCabe, J.T. and Hart, T.C., 1985 Energy extraction and use in a nomadic pastoral ecosystem. *Science* 230: 619-25.

Cousins, B. 1987 A survey of current grazing schemes in the communal lands of Zimbabwe. Mimeo. Harare: Center for Applied Social Sciences, University of Zimbabwe.

Cowles, H.C. 1899 The ecological relations of the vegetation on the sand dunes of Lake Michigan *Botanical Gazette* 27: 97-117, 167-202, 281-308, 361-391.

Cowlishaw, S.J. 1969 The carrying capacity of pastures. *Journal of the British Grassland Society* 24: 207-214.

Cumming, D.H.M. 1982 The influence of large herbivores on savanna structure in Africa. In Huntley, B.J. and Walker, B.H. (eds.).

Dahl, G. and Hjort, A. 1976 *Having Herds*. Stockholm: Department of Social Anthropology, University of Stockholm.

Danckwerts, J.E. 1990 The use of fire to provide nutritious grazing for livestock. *Grassland Society of Southern Africa Bulletin*, press 1.

Danckwerts, J.E. and Stuart-Hill, G.C. 1988 The effect of severe drought and management after drought on the mortality and recovery of semi-arid grassveld. *Journal of the Grassland Society of Southern Africa* 5(4): 218-222.

Danckwerts, J.E. and King, P.G. 1984 Conservative stocking or maximum profit: a grazing management dilemma? *Journal of the Grassland Society of Southern Africa* 1(4): 25-28.

Dancy K.J., Webster, R., and Abel, N.O.J., 1986 Estimating and mapping grass cover and biomass from low-level photographic sampling. *International Journal of Remote Sensing* 712: 1679-1704.

Darwin, C. 1859 *On the Origin of Species*. London: John Murray.

Dasmann, R.F. 1981 *Wildlife Biology*. New York: John Wiley & Sons.

Dasmann, W. 1945 A method for estimating carrying capacity of range lands. *Journal of Forestry* 43: 400-402.

de Haan, C. 1990 Changing trends in the World Bank's lending programme for rangeland development. In Cincotta, R. (ed.) *Suitable Yield System: Implications for the World's Rangelands*. Logan, Utah: Range Science Department, Utah State University.

de Leeuw, P.N. and de Haan, C. 1983 A proposal for development in the pastoral zone in the Republic of Niger. *Pastoral Systems Research in Sub-Saharan Africa*: 435-461. Addis Ababa: IDRC (International Development Research Centre)/ILCA (International Livestock Centre for Africa).

de Leeuw, P.N., Bekure, S. and Grandin, B.E. 1984 Aspects of livestock productivity in Maasai group ranches in Kenya. *ILCA Bulletin* 19: 17-20.

de Leeuw, P.N. and Nyambaka, R. 1988 The prediction of rangeland primary production from rainfall data in arid and semi-arid Eastern Africa. In Dzowela, B.H. (ed.) *Proceedings of the Third PANESA Workshop*: 260-268. Addis Ababa: ILCA (International Livestock Centre for Africa).

de Leeuw, P.N. and Milligan, K. 1983 Resources animales. In Wilson, R.T., de Leeuw, P.N. and de Haan, C. (eds.) *Recherches sur les systèmes des zones arides du Mali: Résultats préliminaires,* p. 51-59. Rapport de Recherche No. 5. Addis Ababa: ILCA (International Livestock Centre for Africa).

de Leeuw, P.N. and Tothill, J.C. 1990 The concept of rangeland carrying capacity in sub-Saharan Africa - myth or reality. *Pastoral Development Network* Paper 29b. London: Overseas Development Institute (also Chapter in this book).

de Ridder, N. and Wagenaar, K.T. 1984 A comparison between the productivity of traditional livestock systems and ranching in eastern Botswana. *ILCA Newsletter* 33:5-7. Addis Ababa: International Livestock Centre for Africa.

de Ridder, N. and Wagenaar, K.T. 1986 Energy and protein balances in traditional livestock systems and ranching in eastern Botswana. *Agricultural Systems* 20: 1-16.

Deshmukh, I.K. 1984 A common relationship between precipitation and grassland peak biomass for east and southern Africa. *African Journal of Ecology* 22: 181-186.

Dhondt, A.A. 1989 Carrying capacity: a confusing concept. *Acta Oecologica* 9: 337-346.

Diarra, L. 1983 *Production et Gestion des Parcours Sahéliens. Synthèse de Cinq Années de Recherches au Ranch de Niono.* Document de Programme. AZ 94. Bamako: International Livestock Centre for Africa.

Dicko-Touré, M. 1980 The contribution of browse to cattle fodder in the sedentary system of the Office du Niger. In Le Houérou, H.N. (ed.) *Browse in Africa* pp. 313-319. Addis Ababa: International Livestock Centre for Africa.

Dillon, J.L. and Anderson, J.R. 1990 *The Analysis of Response in Crop and Livestock Production.* Sydney: Pergamon Press.

Donaldson, T.J. 1986 *Pastoralism and Drought: A Case Study of the Borana of Southern Ethiopia.* M.Phil. thesis, Department of Agriculture and Horticulture, University of Reading, Reading.

Donnelly, J.R. 1988 Technology and the rural industries, expert systems and farm management. pp. 135-139 *Proceedings of Bicentennial Electrical Engineering Congress.* Melbourne.

Doorenbos, J. and Kassam, A. 1979 Yield response to water. *Irrigation and Drainage Paper* No. 33. Rome: FAO (Food and Agriculture Organisation of the United Nations).

Dreyfus, B.L. and Dommergues, Y.R. 1981 Innoculation of *Acacia* species by fast- and slow-growing tropical strains of rhizobium. *Applied Environmental Microbiology* 41: 97-99.

Drury, W.H. and Nisbet, I.C.T. 1973 Succession. *Journal Arnold Arboretum* (Harvard University) 54: 331-368.

Dublin, H.T., Sinclair, A.R.E., Boutin, S., Anderson, E., Jago, M. and Arcese, P., 1990 Does competition regulate ungulate populations? Further evidence from the Serengeti, Tanzania. *Oecolgoia* 82: 283-88.

Duivenbooden, N. van and Gosseye, P.A. (eds.) 1990 Compétition pour des ressources limitées: Le cas de la cinquème région du Mali. Rapport 2: *Productions Végétables, Animales et Hallientiques.* Wageningen: CABO and Mopti, Mali: ESPR.

Dunne, T., Dietrich, W. and Brunengo, M. 1979 Rapid evaluation of soil erosion and soil lifespan in the grazing lands of Kenya. In *IAHS: The Hydrology of Areas of Low Precipitation.* IAHS Publication No. 128.

Dupire, M. 1962 Trade and markets in the economy of the nomadic Fulani of Niger Bororo. In Bohannan, P. and Dalton, G. (eds.) *Markets in Africa*. Chicago: Northwestern University Press.

Dye, P.J. 1983 *Prediction of Variation in Grass Growth in a Semi-arid Induced Grassland*. Ph.D. thesis, Faculty of Science, University of Witswatersrand, Johannesburg.

Dye, P.J. and Spear, P.T. 1982 The effects of bush clearing and rainfall on grass yield and composition in south west Zimbabwe. *Zimbabwe Journal of Agricultural Research* 20: 103-118.

Dyson-Hudson, R. 1972 Pastoralism: Self-image and behavioural reality. In Irons, W. and Dyson-Hudson, N. (eds.) *Perspectives on Nomadism*. Leiden: Brill.

East, R. 1984 Rainfall, soil nutrient status and biomass of large African mammals. *African Journal of Ecology* 22: 245-270.

Eldredge, N., and Gould, S.J. 1972 Punctuated equilibria: An alternative to phyletic gradualism. In Schopf, T.J.M. (ed.) *Models in Paleobiology*. San Francisco: Freeman, Cooper.

Edwards, R.Y. and Fowle, C.D. 1955 The concept of carrying capacity. In Trefethen, J.B. (ed.) *Transactions of the Twentieth North American Wildlife Conference*. pp. 589-602. Washington, DC.: Wildlife Management Institute.

Ellis, J.E. 1991 The role of wildlife production systems in animal agriculture: Sub-Saharan Africa. In *Assessment of animal agriculture in sub-Saharan Africa*. Morrilton, Arkansas: Winrock International.

Ellis, J.E, Galvin, K., McCabe, J.T. and Swift, D.M. 1987 *Pastoralism and Drought in Turkana District, Kenya* A report to NORAD (Norwegian Agency for International Development), Nairobi, Kenya.

Ellis J.E. and Swift, D.M. 1988 Stability of African pastoral ecosystems: Alternate paradigms and implications for development. *Journal of Range Management* 41: 450-459.

Elwell, H. and Stocking, M.A. 1982 Developing a simple, yet practical, method of soil loss estimation. *Tropical Agriculture* (Trinidad) 591: 43-48.

Elwell, H. and Stocking, M.A. 1984 Estimating soil lifespan for conservation planning. *Tropical Agriculture* (Trinidad) 612: 148 - 150.

Evangelou, P. 1984 *Livestock Development in Kenya's Maasailand: Pastoralists' Transition to a Market Economy*. Boulder, Colorado: Westview Press.

FAO (Food and Agriculture Organization of the United Nations) 1988 Guidelines: land evaluation for extensive grazing. *Soil Bulletin* No. 58. Rome: FAO.

Field, D. 1977 *Potential Carrying Capacity of Rangeland in Botswana*. Gaborone: Land Utilization Division, Department of Agricultural Field Services, Ministry of Agriculture.

Field, D. 1980 *The measurement of rangeland resources in Somalia*. Extension Pamphlet No. 6. Mogadishu: National Range Agency.

Flannery, T.F. 1990 Pleistocene faunal loss: implications of the aftershock for Australia's past and future. *Archaeology in Oceania* 25: 45-67.

Flint, M.E.S. 1986 *Crop and Livestock Production in the Pelotshetlha Lands Area*. Main Report of the 1982/83 IFPP Farm Management Survey. Gaborone: Integrated Farming Pilot Project, Ministry of Agriculture.

Foran, B.D. 1986 The impact of rabbits and cattle on an arid calcareous shrubby grassland in central Australia. *Vegetatio* 66: 47-59.

Foran, B.D., Bastin, G. and Shaw, K.A. 1986 Range assessment and monitoring in arid lands: the use of classification and ordination in range survey. *Journal of Environmental Management* 22: 67-84.

Foran, B.D. and Stafford Smith, D.M. 1991 Risk, biology and drought management strategies for cattle stations in central Australia. *Journal of Environmental Management* 33: 17-33.

Fowler, C. 1981 Density dependence as related to life history strategy. *Ecology* 62: 602-610.

Fowler, M.G. 1981 Overgrazing in Swaziland: a review of the technical efficiency of the herd. *Pastoral Network* Paper No.12d. London: Overseas Development Institute.

Franklin, N.N. 1948 *Economics in South Africa*. Cape Town: Oxford University Press.

Friedel, M.H. 1990 Some key concepts for measuring Australia's arid and semi-arid rangelands. *Australian Rangeland Journal* 12: 21-24.

Friedel, M.H. 1991 Range condition assessment and the concept of thresholds: A viewpoint. *Journal of Range Management* 44(5): 422-433.

Friedel, M.H., Pickup, G. and Nelson D.J. 1993 The interpretation of vegetation change in a spatially and temporally diverse arid Australian landscape. *Journal of Arid Environments* 24: in press.

Frost, P., Medina, E., Menaut, J-C., Solbrig, O., Swift, M. and Walker, B. 1986 Responses of savannas to stress and distrubance: A proposal for a collaborative programme of research. Report of IUBS working group on Decade of Tropics/Tropical savanna ecosystems. Special Issue 10, *Biology International,* Paris: IUBS (International Union of Biological Sciences).

Fry, P.H. and McCabe, J.T. 1986 A comparison of the survey methods on pastoral Turkana migration patterns and the implications for development planning. *Pastoral Development Network* No. 22b. London: Overseas Development Institute.

Gallais, J. 1975 *Pasteurs et Paysans de Gourma.* Paris: Centre National de la Recherche Scientifique.

Gates, P.J. and Brown, K. 1988 *Acacia tortilis* and *Prosopis cineraria*: Leguminous trees for arid areas. *Outlook on Agriculture* 17: 61-64.

Giles, R.H. 1978 *Wildlife management.* San Francisco: W.H. Freeman.

Gleason, H.A. 1927 Further views on the succession concept. *Ecology* 8: 299-326.

Goldstein, M.C., Beall, C.M. and Cincotta, R.P. 1990 Traditional conservation on Tibet's northern plateau. *National Geographic Research* 62: 139-56.

Goodall, D.W. 1967 Computer simulation of changes in vegetation subject to grazing. *Journal of the Indian Botanical Society* 46: 356-362.

Government of Zimbabwe 1986 *First Five Year Development Plan, 1986-1990.* Harare: Government Printer.

Government of Zimbabwe 1987 *The National Conservation Strategy.* Harare: Department of Information, Ministry of Natural Resources and Tourism.

Grandin, B. 1987 Pastoral culture and range management: Recent lessons from Maasailand. *ILCA Bulletin* 28: 7-13.

Griffin, G.F. and Friedel, M.H. 1984 Effects of fire on central Australian rangelands: Changes in tree and shrub populations. *Australian Journal of Ecology* 9: 395-403.

Griffin, G.F. and Friedel, M.H. 1985 Discontinuous change in central Australia: Some implications of major ecological events for land management. *Journal of Arid Environments* 9: 63-80.

Grouzis, M. 1990 *Dynamics of Sahelian Ecological Systems: The Case of Oursi Pond, Burkina Faso.* Paper delivered at Commonwealth Secretariat Technical Meeting on Savanna Development and Pasture Production, Woburn, UK, Nov. 1990.

Grouzis, M. and Sicot, M. 1983 Production frequentielle du bassin versant de la mare d'Oursi: application à l'estimation des potentialités pastorales. In Vanpraet, C. (ed.) *Méthodes d'Inventaire et de Surveillance Continue des Ecosystemes Pastoraux Saheliens.* pp. 177-188. Dakar: ISRA/FAO (Food and Agriculture Organisation of the United Nations).

Hacker R. 1990 *Range condition and shrub encroachment in the SORDU pilot project area.* Report commissioned by the Fourth Livestock Development Project of the Ethiopian Ministry of Agriculture. GRM International Pty Ltd.

Hacker, R., Beurle,D. and Gardiner, G. 1990 Monitoring Western Australia's rangelands. *Western Australian Journal of Agriculture* 31: 33-38.

Hacker, R.B, Wang, K-M., Richmond, G.S. and Lindner, R.K. 1991 IMAGES: An integrated model of an arid grazing ecological system. *Agricultural Systems* 37: 119-163.

Hadwen, S. and Palmer, L.J. 1922 Reindeer in Alaska. *USDA Bulletin* No. 1089. Washington, DC: US Department of Agriculture.

Hanson, J.D., Skiles, J.W. and Parton, W.J. 1988 A multi-species model for rangeland plant communities. *Ecological Modelling* 44: 89-123.

Hardin, G. 1968 The tragedy of the commons. *Science* 162: 1243-1248.

Hardy, M.B., and Hurt, C.R. 1989 An evaluation of veld condition assessment techniques in Highland Sourveld. *Journal of the Grassland Society of Southern Africa* 6(2): 51-58.

Harper, J.L. 1977 *Population Biology of Plants*. New York: Academic Press.

Harrington, G.N., Wilson, A.D. and Young M.D., 1984 *Management of Australia's Rangelands*. Melbourne: Division of Wildlife and Rangelands Research, CSIRO (Commonwealth Scientific and Industrial Research Organisation).

Harris, P. S. and O'Connor, K.F. 1980 The grazing behaviour of sheep *Ovis aries* on a high-country summer range in Canterbury, New Zealand. *New Zealand Journal of Ecology* 3: 85-96.

Hart, R.H. 1972 Forage yield, stocking rate, and beef gains on pasture. *Herbage Abstracts* 42: 345-353.

Hart, R.H. 1978 Stocking rate theory and its application to grazing on rangelands. pp. 547-550 In Hyder, D., (ed.) *Proceedings of the First International Rangeland Congress*. Colorado: Society for Rangeland Management.

Hassell, M. and May, R. 1985 From individual behaviour to population dynamics. In Sibly, R, and Smith, R. (eds.) *Behavioural Ecology*. Oxford: Blackwell Scientific Publications.

Hassell, M., Southwood, T. and Reader, P. 1987 The dynamics of the viburnum whitefly: a case study on population regulation. *Journal of Animal Ecology* 56: 283-300.

Hatton J.C. and Smart N.O.E. 1984 The effect of long-term exclusion of large herbivores on soil nutrient status in Murchison Falls National Park, Uganda. *African Journal of Ecology* 22: 23-30.

Heady, H.F. 1975 *Rangeland Management*. New York: McGraw-Hill.

Hesse, C., Thiam, A., Fowler, C. and Swift, J. 1984 A Fulani agro-pastoral production system in the Malian Gourma. Draft report. Bamako, Mali: ILCA (International Livestock Centre for Africa).

Hiernaux, P. 1982 Méthode d'évaluation du potential fourrager de parcour sahélien. *Miscellaneous Paper* No 430. Wageningen: CABO (Centre for Agrobiological Research).

Hiernaux, P. 1984 Distribution des pluies et production herbacée au Sahel: une méthode empirique pour caractériser la distribution de précipitations journalières et ses effets sur la production herbacées. *Document de Programme* No. AZ 98. Bamako, Mali: CIPEA/ILCA (International Livestock Centre for Africa).

Hiernaux, P. 1989 Note sur l'évolution de la biomasse des pailles au cours de la saison sèche. Working document. Bamako, Mali: CIPEA/ILCA (International Livestock Centre for Africa).

Hiernaux, P. and Cissé, M.I. 1983 Les ressources naturelles de l'élevage du Gourma Malien. *Document de Programme* AZ 926. Bamako, Mali: CIPEA/ILCA (International Livestock Centre for Africa).

Hiernaux, P., Diarra, L. and Maiga, A. 1990 Dynamique de la végétation sahelienne après la sécheresse. Working document 001/90. Bamako, Mali: ILCA (International Livestock Centre for Africa).

Hiernaux, P. and Justice, C.O. 1986 Suivi du développement végétal au cours de l'été 1984 dans le Sahel Malien. *International Journal of Remote Sensing* 7: 1515-1532.

Higgins 1967 Unreferenced. *Bulldust* 1985: Student Magazine of the Faculty of Agriculture, Natal University.

Hodgkinson, K.C. 1991 Identification of critical thresholds in opportunistic management of rangeland vegetation. *Proceedings of 4th International Rangelands Congress*. Montpellier, France. In press.

Hodgkinson, K.C. and Harrington, G.N. 1985 The case for prescribed burning to control shrubs in eastern semiarid woodlands. *Australian Rangeland Journal* 7: 64-74.

Hodgson, J. 1979 Nomenclature and definitions in grazing studies. *Grass and Forage Science* 34: 11-18.

Holechek, J.L., Pieper, R.D. and Herbel, C.H. 1989 *Range Management, Principles and Practices*. Englewood Cliffs, New Jersey: Prentice Hall.

Holling, C.S. 1973 Resilience and stability of ecological systems *Annual Review of Ecology and Systematics* 4: 1-23.

Holmes, W. 1988 *BREEDCOW and DYNAMA Herd Budgeting Spreadsheet Models.* Townsville, Australia: Department of Primary Industries.

Horn, H.S. 1981 Succession. In May, R.M. (ed.) *Theoretical Ecology: Principles and Applications.* Oxford: Blackwells.

Horowitz, M. and Little, P. 1987 African pastoralism and poverty: Some implications for drought and famine. In Glantz, M. (ed.) *Drought and Hunger in Africa: Denying Famine a Future.* Cambridge: Cambridge University Press.

Hubbard, M. 1986 *Agricultural Exports and Economic Growth. A Study of the Botswana Beef Industry.* London: KPI.

Huntley, B.J. 1982 Southern African savannas. In Huntley, B.J. and Walker, B.H. (eds.) *Ecology of Tropical Savannas.* Berlin: Springer Verlag.

Huntley, B.J. and Walker, B.H. (eds.) 1982 *Ecology of Tropical Savannas.* Ecological studies No. 42, Berlin: Springer-Verlag.

Hutchinson, G.E. 1978 *An Introduction to Population Ecology.* New Haven, CT: Yale University Press.

Hyder, D.N., Bement, R.E., Remmega, E.E., and Terwilliger, C. Jr. 1966 Vegetation soils and vegetation grazing relationships frequency data. *Journal of Range Management* 19: 11-16.

IEMVT (Institut Élevage Médicine Vétérinaire Pays Tropicaux) 1989 Élevage et Potentialités Pastorales Sahéliennes: Synthèses Cartographiques, vol. Mali. Maisons Alfort, France: IEMVT/CTA (Technical Centre for Agricultural and Rural Cooperation).

Jahnke H.E. 1982 *Livestock Production Systems and Livestock Development in Tropical Africa.* Kiel: Kieler Wissenschaftsverlag Vauk.

Jarman, P. and Sinclair, A. 1979 Feeding strategy and the pattern of resource partitioning in ungulates. In Sinclair, A. and Norton-Griffiths, M. (eds.) *Serengeti: Dynamics of an Ecosystem.* Chicago: University of Chicago Press.

Jarvis, L.S. 1984 *Overgrazing and Range Degradation in Africa: The Need and Scope for Government Control of Livestock Numbers?* Paper presented to the Conference on Livestock Policy Issues in Africa, International Livestock Centre for Africa, September 1984, Addis Ababa.

Jarvis, L.S. and Erickson, R. 1986 Livestock herds, overgrazing and range degradation in Zimbabwe: How and why do the herds keep growing? *ALPAN Network Paper.* Addis Ababa: ILCA (International Livestock Centre for Africa).

JEPPS (Joint Ethiopian Pastoral Systems Study) 1984 *Interim Research Reports.* Addis Ababa: ILCA (International Livestock Centre for Africa).

Johns, G.G., Tongway, D.J. and Pickup, G. 1984 Land and water processes. In Harrington et al. (eds.).

Johnson, D. (ed.) 1979 *A Report from the Workshop on Pastoralism and African Livestock Production.* Mimeo. Binghamton, New York: Institute for Development Anthropology.

Jones, R.J. and Sandland, R.L. 1974 The relation between animal gain and stocking rate: Derivation of the relation from the results of grazing trials. *Journal of Agricultural Science* 83: 335-341.

Kamara, C.S. and Haque, I. 1988 *Characteristics of Upland Soils at ILCA Research Sites in Ethiopia.* Plant Science Division Working Paper No. B6. Addis Ababa: ILCA (International Livestock Centre for Africa).

Kelly, R.D. 1973 *A Comparative Study of Primary Productivity Under Different Kinds of Land Use in South-eastern Rhodesia.* Ph.D. thesis, University of London.

Kelly, R.D. and Walker, B.H. 1976 The effects of different forms of land-use on the ecology of a semi-arid region in south-eastern Rhodesia. *Journal of Ecology* 64: 553-576.

Kerven, C. (ed.) 1982 *Migration in Botswana: Patterns, Causes and Consequences. Final Report.* National Migration Study, Vol III. Gaborone: Central Statistics Office, Ministry of Finance and Development Planning.

Kirkby, M.J., Atkinson, K. and Lockwood, J. 1990 Aspect, vegetation cover and erosion on semi-arid hillslopes. In Thornes, J.B. (ed.) *Vegetation and Erosion*, pp.25-39. Chichester: John Wiley & Sons.

Klein, H.D. 1981 Contribution à l'estimation de la production sur pâturage sahélien au Niger. *Revue d'Élevage et de Médicine vétérinaire des Pays Tropicaux* 34(2): 211-220.

Knoop, W.T. and Walker, B.H. 1985 Interactions of woody and herbaceous vegetation in a southern African savanna. *Journal of Ecology* 73: 235-253.

Konandreas, P.A., and Anderson, F.M., 1982 *Cattle Herd Dynamics: an Integer and Stochastic Model for Evaluating Production Alternatives.* ILCA Research Report No 2. Addis Ababa: International Livestock Centre for Africa.

KwaZulu Government 1979 to 1980 *Minister's Policy Speech.* Department of Agriculture and Forestry. Ulundi: Government Printer.

KwaZulu Government 1980 A preliminary land-use plan for KwaZulu. Unpublished report. Ulundi: Department of Agriculture and Forestry.

Lamprey H.F. 1983 Pastoralism yesterday and today: The overgrazing problem. In Bourliere, F. (ed.) *Tropical Savannas*. Ecosystems of the World, vol. 13. Amsterdam: Elsevier Press.

Lane, L.J. and Nearing, M.A. (eds.) 1989 *USDA - Water Erosion Prediction Project: Hillslope Model Documentation.* Washington, DC: US Department of Agriculture, Agricultural Research Service.

Lange, R.T. and Willcocks, M.C. 1980 Experiments on the capacity of present sheep flocks to extinguish some tree populations of the South Australian arid zone. *Journal of Arid Environments* 3: 223-230.

Lauenroth, W.K. and Laycock, W.A. 1989 *Secondary Succession and the Evaluation of Range Condition*. Boulder, CO: Westview Press.

Laycock, W.A. 1991 Stable states and thresholds of range condition on North American rangelands: A viewpoint. *Journal of Range Management* 44(5): 427-733.

Le Houérou, H.N. 1989 *The Grazing Land Ecosystems of the African Sahel*. Berlin: Springer-Verlag.

Le Houérou, H.N., Bingham, R.L. and Skerbek, 1988 Relationships between the variability of primary production and the variability of annual precipitation in world arid lands. *Journal of Arid Environments* 15: 1-8.

Le Houérou, H.N. and Hoste, C.H. 1977 Rangeland production and annual rainfall relations in the Mediterranean basin and in the African Sahelo-Sudanian zone. *Journal of Range Management* 30: 181-189.

Leloup, S. and Traoré, M. 1989 La situation fourragère dans le Sud-Est du Mali. Une étude agro-écologique. *DRSPR - Volet Fonsébougou, Sikasso*. Amsterdam: KIT.

Livingstone, I. 1991 Livestock management and 'overgrazing' among pastoralists. *Ambio* 20(2): 80-85.

Ludwig, J. 1990 SHRUBKILL: a decision support system for management burns in Australian savannas. *Journal of Biogeography* 17: 547-550.

Mabbut, J.A. 1985 Desertification of the world's rangelands. *Desertification Control Bulletin* 12: 1-11.

MacDonald, I.A.W. 1985 The Australian contribution to southern Africa's invasive alien flora: an ecological analysis. *Proceedings of the Ecological Society of Australia* 14: 225-236.

Mace, R. and Houston, A. 1989 Pastoralist strategies for survival in unpredictable environments: A model of herd composition that maximises household viability. *Agricultural Systems* 31: 185-204.

Mack, R. 1970 The great African cattle plague epidemic of the 1890s. *Tropical Animal Health and Production* 2: 210-219.

Macnab, J. 1985 Carrying capacity and related slippery shibboleths. *Wildlife Society Bulletin* 13: 403-410.

MacTainsh, G.H. 1989 Quaternary aeolian dust processes and sedimentation in the Australian region. *Quaternary Science Review* 8: 235-253.

Maiga, A., Leeuw, P.N. de, Diarra, L. and Hiernaux, P. 1991 La récolte des céréales sauvages dans la Région du Gourma au Mali. *Programme Reseaux des Zones Arides* Dossier 27. London: IIED (International Institute for Environment and Development).

Malthus, T. 1798 *Essay on the Principle of Population*. Reprinted 1976. New York: Phillip Appleman ed. Norton.

Margalef, R. 1968 *Perspectives in Ecological Theory*. Chicago: University of Chicago Press.

May, R. 1977 Thresholds and breakpoints in ecosystems with a multiplicity of stable states. *Nature* 269: 471-477.

May, R. 1989a Levels of organisation in ecology. In Cherret, J. (ed.) *Ecological Concepts: The Contribution of Ecology to an Understanding of the Natural World*. Oxford: Blackwell Scientific.

May, R. 1989b The chaotic rhythms of life. *New Scientist* 18 (November): 37-41.

McCabe, J.T. 1987a Drought and recovery: Livestock dynamics among the Ngisonyoka Turkana of Kenya. *Human Ecology* 15: 317-329.

McCabe, J.T. 1987b Inter-household variation in Ngisonyoka Turkana livestock production. *Research in Economic Anthropology* 8: 277-93.

McKay, A.D. 1968 Rangeland and productivity in Botswana. *East African Agriculture and Forestry Journal* 34: 178-192.

McKenzie, B. 1982 Resilience and stability of the grasslands of Transkei. *Proceedings of the Grassland Society of Southern Africa* 17: 21-24.

McKeon, G.M., Day, K.A., Howden, S.M., Mott, J.J., Orr, D.M., Scattini, W.J. and Weston, E.J. 1990 Northern Australian savannas: management for pastoral production. *Journal of Biogeography* 17: 355-372.

McKeon, G.M. and Rickert, K.J. 1984 A computer model of the integration of forage options for beef production. *Proceedings of the Australian Society of Animal Production* 15: 15-19.

McKeon, G.M., Rickert, K.J., Ash, A.J., Cooksley, D.G. and Scattini, W.J. 1982 Pasture production model. *Proceedings of the Australian Society of Animal Production* 14: 201-204.

McNaughton, S.J. 1985 Ecology of a grazing ecosystem: The Serengeti. *Ecological Monographs* 553: 259-94.

Meadows S.J. and White J.M. 1979 Structure of the herd and determinants of offtake rates in Kajiado District in Kenya 1962-1977. *Pastoral Development Network* Paper No. 11e. London: Overseas Development Institute.

Mentis, M.T. 1985 Conserving rangeland resources. *Journal of the Grassland Society of Southern Africa* 2: 27-31.

Mentis, M.T., Grossman, D., Hardy, M.B., O'Connor, T.G. and O'Reagain, P.J. 1989 Paradigm shifts in South African range science, management and administration. *South African Journal of Science* 85: 684-687.

Miles, R.L. and Johnston P.W. 1990 Run-off and soil loss from four small catchments in the mulga lands of south-west Queensland. *Proceedings of the 6th Australian Rangeland Conference* pp. 170-184. Carnarvon, Western Australia.

Milligan, K. 1982 *Wet Season Aerial Survey of the Human and Livestock Populations and Environmental Condition in a Central Region of the Pastoral Zone of Niger*. Niamey, Niger: USAID (United States Agency for International Development)/CIPEA (International Livestock Centre for Africa).

Milligan, K. 1983a *An Aerial Reconnaissance of Livestock and Human Population in Relation to Land Use and Ecological Conditions in the SORDU Project Area of Southern Ethiopia*. JEPSS Joint Ethiopian Pastoral Systems Study Research Report No. 5. Addis Ababa: ILCA (International Livestock Centre for Africa).

Milligan, K. 1983b *Dry Season Patterns of Livestock and Human Distribution in the Gourma Region of Mali*. Addis Ababa: ILCA (International Livestock Centre for Africa).

Milne, G. 1947 A soil reconnaissance journey through parts of Tanganyika territory. *Journal of Ecology* 35: 192-265.

Milton, S.J. 1991 Plant spinescence in arid southern Africa: Does moisture mediate selection by mammals? *Oecologia* 87: 279-287.

Milton, S.J. and Dean, W.R.J. 1990 Seed production in rangelands of the southern Karoo. *South African Journal of Science* 86: 231-233.

Moore, A.D. 1990 The semi-Markov process: a useful tool in the analysis of vegetation dynamics for management. *Journal of Environmental Management* 30: 111-130.

Morgan, R.P.C. 1979 *Soil Erosion*. New York: Longman.

Morison, C., Hoyle, A. and Hope-Simpson, J. 1948 Tropical soil vegetation catenas and mosaics: A study in south western part of Anglo-Egyptian Sudan. *Journal of Ecology* 36: 1-48.

Morton, S.R. and Andrew, M.H. 1987 Ecological impact and management of fire in Northern Australia. *Search* 18: 77-82.

Mott, G.O. 1960 Grazing pressure and the measurement of pasture production. In *Proceedings of the 8th International Grassland Congress* pp. 606-611. Reading, England.

Mott, J.J., Ludlow, M.M., Richards, J.H. and Parsons A.R. 1992 Effects of moisture supply in the dry season and subsequent defoliation on persistance of the savanna grasses *Themeda triandra, Heteropogon contortus,* and *Panicum maximum. Australian Journal of Agricultural Research* 43: 241-260.

Msafiri, D.N. and Pieper R.D. 1989 Vegetational response to herbicide treatment for brush control in Tanzania. *Journal of Range Management* 42: 332-336.

Namken, J.C. and Stuth, J.W. 1991 A prototype spatial algorithm for predicting grazing distribution pattern by cattle. In Stuth, J.W. (ed.) *Proceedings of International Conference on Decision Support Systems for Resource Management,* pp. 77-81. College Station, Texas: Texas A&M University.

Nestel P.S. 1985 *Nutrition of Maaasai Women and Children in Relation to Subsistence Food Production*. Ph.D. thesis, Nutrition Department, Queen Elizabeth College, University of London.

Newsome, A.E., Dndzinski, M.L. and Low, W.A. 1981 Measuring bias in aerial surveys due to dispersion of animals, pp.185-194 in *Low-level Aerial Survey Techniques: Report of Nairobi Workshop*. Addis Ababa, Ethiopia: International Livestock Centre for Africa (ILCA).

Nicholls, N. and Wong, K.K. 1990 Dependence of rainfall variability on mean rainfall, latitude, and the Southern Oscillation. *Journal of Climate* 3: 163-170.

Nicholson, A. 1933 The balance of animal populations. *Journal of Animal Ecology* 2: 132-178.

Nicholson, S.E. 1980 The nature of rainfall fluctuations in sub-tropical West Africa. *Monthly Weather Review* 108: 473-487.

Nicholson, S.E. 1986 The nature of rainfall variability in Africa south of the Equator. *Journal of Climatology* 6: 515-30.

Noble, I.R. 1975 *Computer simulations of sheep grazing in the arid zone*. Ph.D. thesis, University of Adelaide.

Noble, I.R. and Slatyer R.O. 1980 The use of vital attributes to predict successional changes in plant communities subject to recurrent disturbance. *Vegetatio* 43: 5-21.

Noble, J.C. 1989 Fire studies in mallee *Eucalyptus* spp. communities of western New South Wales: The effects of fires applied in different seasons on herbage productivity and their implications for management. *Australian Journal of Ecology* 14: 169-187.

Norton-Grifffiths, M. 1979 The influence of grazing, browsing, and fire on the vegetation dynamics of the Serengeti. In Sinclair, A.R.E. and Norton-Griffiths, M. (eds.) *Serengeti: Dynamics of an Ecosystem*. Chicago: University of Chicago Press.

Norval, R. 1978 The effects of a partial breakdown of dipping in African areas of Rhodesia. *Rhodesian Veterinary Journal* 9: 9-16.

Noy-Meir, I. 1973 Desert ecosystems: Environment and producers. *Annual Review of Ecology and Systematics* 4: 25-51.

Noy-Meir, I. 1975 Stability of grazing systems: An application of predator-prey graphs. *Journal of Ecology* 63: 459-481.

Noy-Meir, I. and Walker, B.H. 1986 Stability and resilience in rangelands. In Joss, P.J., Lynch, P.W. and Williams, O.B. (eds.) *Rangelands: A Resource under Siege,* pp. 21-25. Canberra: Australian Academy of Sciences.

O'Connor, T.G. 1985 A synthesis of field experiments concerning the grass layer in the savanna regions of southern Africa. *South African National Scientific Programme Report* No. 114. Pretoria, South Africa: Council for Scientific and Industrial Research.

O'Reagain, P.J. and Mentis, M.T. 1990 The effect of veld condition on the quality of diet selected by cattle grazing the Natal Sour Sandveld. *Journal of the Grassland Society of South Africa* 7: 190-195.

ODI (Overseas Development Institute) 1977 United Nations Conference on Desertification. *Briefing Paper.* London: ODI.

Odum, E.P. 1953 *Fundamentals of Ecology*. Philadelphia: W.B. Saunders.

Odum, E.P. 1969 The strategy of ecosystem development. *Science* 164: 262-70.

Odum, E.P. 1983 *Basic Ecology*. Philadelphia: Saunders College Publishing.

Parsons, J. 1986 Sediment yields in five Eastern Transvaal catchments. Mimeo. Silverton, Republic of South Africa: Division of Agricultural Engineering.

Parton, W. J., Cole, C.V., Stewart, J.W.B., Schimel, D.S. and Ojima, D.S. 1989 Simulating the long-term dynamics of C, N and P in soils. In Clarholm, M. and Bergstrom, L. (eds.) *Ecology of Arable Land - Perspectives and Challenges, Developments in Plant and Soil Sciences* Vol. 39. Kluwer Academic Publishers.

Payne, W.J.A. 1965 Specific problems of semi-arid environments. *Qualitas Plantarum et Materiae Vegetabiles* 12 (3): 269-294.

Payne, W.J.A. 1990 *Rangeland Management: The Significance of the Concept of Carrying Capacity*. Paper delivered at Commonwealth Secretariat Technical Meeting on Savanna Development and Pasture Production, Woburn, UK, Nov. 1990.

Pearl, R. and Reed, L.J. 1920 On the rate of growth of the population of the United States since 1790 and its mathematical representation. *Proceedings of the National Academy of Science* 6: 275-288.

Peco, B. 1989 Modelling Mediterranean pasture dynamics. *Vegetatio* 83: 269-276.

Penning de Vries, F.W.T. and Djitèye, M.A. (eds.) 1982 *La Productivité des Pâturages Sahéliens: Une Étude des Sols, des Végétations et de l'Exploitation de cette Ressource Naturelle.* Report No. 918. Wageningen: Centre for Agricultural Publishing and Documentation.

Pentz, J.A. 1955 Grasses in soil and water conservation. In Meredith, D. (ed.) The Grasses and Pastures of South Africa.

Pickup, G. 1985 The erosion cell - a geomorphic approach to landscape classification in range assessment. *Australian Rangeland Journal* 7: 114-121.

Pickup, G. 1988 Modelling soil erosion at the regional scale. In Warner, R.F. (ed.) *Fluvial Geomorphology of Australia*, pp. 105-127. Sydney: Academic Press.

Pickup, G. 1989 New land degradation survey techniques for arid Australia - problems and prospects. *Australian Rangeland Journal* 11: 74-82.

Pickup, G. 1991 Event frequency and landscape stability on the floodplain systems of arid central Australia. *Quaternary Science Reviews* 10: 463-473.

Pickup, G. 1993 Modelling patterns of defoliation by grazing animals in rangelands. Submitted to *Journal of Applied Ecology*.

Pickup, G. and Chewings, V.H. 1986 Random field modelling of spatial variations in erosion and deposition on flat alluvial landscapes in arid Central Australia. *Ecological Modelling* 33: 269-296.

Pickup, G. and Chewings, V.H. 1988a Forecasting patterns of erosion in arid lands from Landsat MSS data. *International Journal of Remote Sensing* 9: 69-84.

Pickup, G. and Chewings, V.H. 1988b Estimating the distribution of grazing and patterns of cattle movement in a large arid zone paddock - an approach using animal distribution models and Landsat imagery. *International Journal of Remote Sensing* 9: 1469-1490.

Pickup, G and Chewings, V.H. 1992 A grazing gradient approach to land degradation assessment in arid areas from remotely-sensed data. *International Journal of Remote Sensing*, in press.

Pickup, G. and Stafford Smith, D.M. 1993 Problems, prospects and procedures for assessing sustainability of pastoral land management in arid Australia. Submitted to *Journal of Biogeography*.

PIRT 1983 *Projet Inventaire des Ressources Terrestres au Mali.* 3 volumes. New York: Government of Mali/TAMS/USAID (United States Agency for International Development).

Planning and Statistics Division 1978-1988 *Botswana Agricultural Statistics.* Gaborone: Ministry of Agriculture and the Central Statistics Office.

Planning and Statistics Division 1980-1984 *Farm Management Survey Results.* Gaborone: Ministry of Agriculture and the Central Statistics Office.

Powell, J.M. 1986 Manure for cropping: A case study from central Nigeria. *Experimental Agriculture* 22: 15-24.

Pratt, D.J. and Gwynne M.D. (eds.) 1977 *Rangeland Management and Ecology in East Africa.* London: Hodder and Stoughton.

Pratt, D J. 1987a *Status Report on Rangelands and Range Development in the Sidamo and Harerge Provinces of Ethiopia: Draft Main Report.* FAO Project TCP/ETH/4535, commissioned by the Ministry of Agriculture of the People's Democratic Republic of Ethiopia. Rome: FAO (Food and Agriculture Organisation of the United Nations).

Pratt, D.J. 1987b *Status Report on Rangelands and Range Development in the Sidamo and Harerge Provinces of Ethiopia: Draft Report Annex.* FAO Project TCP/ETH/4535, commissioned by the Ministry of Agriculture of the People's Democratic Republic of Ethiopia. Rome: FAO (Food and Agriculture Organisation of the United Nations).

Prescott, J.R.V. 1961 Overpopulation and overstocking in the native areas of Matabeleland. *Geography Journal* 127: 212-225.

Prince, S.D. and Justice, C.O. 1991 Course resolution remote sensing of the Sahelian environment. Special issue of *International Journal of Remote Sensing* 12: 1133-1405.

Prince, S.D., Justice, C.O. and Los, S.O. 1990 *Remote Sensing of the Sahelian Environment.* Brussels: Commission of the European Communities.

Purnell, M.F. 1983 The FAO methods of land evaluation for agriculture and forestry as compared to extensive grazing. In Siderius (ed.), pp. 133-143.

Raay, van H.G.T. and de Leeuw, P.N. 1970 The importance of crop residues as fodder: a resource analysis in Katsina Province, Nigeria. *Tijdschrift voor Econ. en Soc. Geografie* 61: 137-147.

Raay, van H.G.T. and de Leeuw, P.N. 1974 *Fodder Resources and Grazing Management in a Savanna Environment.* Occasional Paper No 45. The Hague: Institute of Social Studies.

Radwanski, S.A. 1969 Improvement of red acid sands by the neem tree *Azadirachta indica* in Sokoto, Northwestern State of Nigeria. *Journal of Applied Ecology* 6: 507-511.

Radwanski, S.A. and Wickens, G.E. 1967 The ecology of *Acacia albida* on mantle soils in Zalingei, Jebel Marra, Sudan. *Journal of Applied Ecology* 4: 569-579.

Rattray, J. 1957 The grasses and grass associations of Southern Rhodesia. *Rhodesian Agriculture Journal* 54: 197-234.

Rennie, T.W., Light, D., Rutherford, A., Miller, M., Fisher, I., Pratchett, D., Capper, B., Buck, N. and Trail, J. 1977 Beef cattle productivity under traditional and improved management in Botswana. *Tropical Animal Health and Production* 9: 1-6.

Ricklefs, R.E. 1979 *Ecology.* New York: Chiron Press.

Riewe, M.E. 1961 Use of the relationship of stocking rate to gain of cattle in an experimental design for grazing trials. *Agronomy Journal* 53: 309-313.

Riney, T. 1982 *Study and Management of Large Mammals.* New York: John Wiley & Sons.

Roberts, C.R. 1980 Effect of stocking rate on tropical pastures. *Tropical Grasslands* 14: 225-231.

Rodriguez, A. and Jameson, D.A. 1988 Rainfall risk in grazing management. *Ecological Modelling* 41: 85-100.

Roy, J.H.B. 1980 *The Calf.* Fourth Edition. London: Butterworths.

Rutherford, M.C. 1978 Primary production ecology in Southern Africa. In Wager, M.J.A. (ed.) *Biogeography and Ecology of Southern Africa*, pp. 623-652. The Hague: Junk.

Sabiiti, E.N. and Wein, R.W. 1988 Fire behaviour and the invasion of *Acacia sieberiana* into savanna grassland openings. *African Journal of Ecology* 26: 301-313.

Sampson, A.W. 1913 Range improvement by deferred and rotation grazing. *USDA Bulletin* No. 34. Washington, DC: US Department of Agriculture.

Sampson, A.W. 1923 *Range and Pasture Management*. New York: John Wiley & Sons.

Sandford, S. 1976 The design and management of pastoral development. *Pastoral Network Paper* No. 1d. London: Overseas Development Institute.

Sandford, S. 1977 *Dealing With Drought and Livestock In Botswana*. London: Overseas Development Institute.

Sandford S. 1979 *Keeping An Eye On Tribal Grazing Lands Policy*. Gaborone, Botswana: National Institute of Research and London: Overseas Development Institute.

Sandford, S. 1982a *Livestock in the Communal Areas of Zimbabwe. A Report to the Ministry of Lands, Resettlement and Rural Development*. London: Overseas Development Institute.

Sandford, S. 1982b Pastoral Strategies and Desertification: Opportunism and Conservatism in Dry Lands. In Spooner B. and Mann, H.S. (eds.) *Desertification and Development* pp. 61-80. London: Academic Press.

Sandford, S. 1983a *Management of Pastoral Development in the Third World*. Chichester: John Wiley & Sons.

Sandford, S. 1983b Organisation and management of water supplies in tropical Africa. *ILCA Research Report* No.8. Addis Ababa: ILCA (International Livestock Centre for Africa).

Scanlan, J.C. and Burrows, W.H. 1990 Woody overstorey impact on herbaceous understorey in *Eucalyptus* spp. communities in central Queensland. *Australian Journal of Ecology* 15: 191-197.

Scarnecchia, D.L. 1990 Concepts of carrying capacity and substitution ratios: A systems viewpoint. *Journal of Range Management* 43: 553-555.

Schlesinger, W.H., Fonteyn, P.J. and Reiners, W.A. 1989 Effects of overland flow on plant water relations, erosion, and soil water percolation on a Mojave Desert landscape. *Soil Science Society of America Journal* 53: 1567-1572.

Scoones, I. 1988 *Sustainable Pastoralism in Africa. An Annotated Bibliography*. London: Drylands Programme, International Institute for Environment and Development.

Scoones, I. 1989a Economic and ecological carrying capacity: Implications for livestock development in Zimbabwe's communal areas. *Pastoral Development Network* Paper No. 27b. London: Overseas Development Insitute.

Scoones, I. 1989b Patch use by cattle in dryland Zimbabwe: farmer knowledge and ecological theory. *Pastoral Development Network* Paper No. 28b. London: Overseas Development Institute.

Scoones, I. 1990 *Livestock Populations and the Household Economy: A Case Study from Southern Zimbabwe* Ph.D. thesis, University of London.

Scoones, I. 1993 Coping with drought: Responses of herders and livestock in contrasting Savanna environments in southern Zimbabwe. *Human Ecology* 20: 31-52.

Scotney, D.N. and Mcphee, P.J. 1990 The dilemma of our soil resources. Paper prepared for National Trust Conference, Pretoria, South Africa.

Scott, J.D. 1981 Soil erosion, its causes and prevention. In Tainton, N.M. (ed.) *Veld and Pasture Management in South Africa*. Pietermaritzburg: Shuter & Shooter.

Seligman, N.G. and van Keulen, H. 1989 Herbage production of a Mediterranean grassland in relation to soil depth, rainfall and nitrogen nutrition: A simulation study. *Ecological Modelling* 47: 303-311.

Senft, R.L., Coughenour, M.B., Bailey, D.W., Rittenhouse, L.R., Sala, O.E. and Swift, D.M. 1987 Large herbivore foraging and ecological hierarchies: Landscape ecology can enhance traditional foraging theory. *BioScience* 37: 789-799.

Senft, R.L., Rittenhouse, L.R. and Woodmansee, R.G. 1983 The use of regression models to predict spatial patterns of cattle behavior. *Journal of Range Management* 36: 553-557.

Sharman, M. J. 1983 Approche systémique du suivi continu d'une zone pastorale. In Vanpraet, C. (ed.) *Mèthodes d'inventaire et de surveillance continue des ecosystemes pastoreaux saheliens*, pp. 149-160. Dakar: ISRA/FAO (Food and Agriculture Organisation of the United Nations).

Shepherd, N. and Caughley, G. 1987 Options for management of kangaroos. In Caughley et al. (eds.).

Shmida, A. and Ellner, S. 1984 Coexistence of plant species with similar niches. *Vegetatio* 58: 29-55.

Siderius, W. (ed.) 1983 Proceedings of the Workshop on Land Evaluation for Extensive Grazing (LEEG). Wageningen: International Institute for Land Reclamation and Improvement.

Sinclair, A. 1989 Population regulation in animals. In Cherret, J. (ed.) *Ecological Concepts: The Contribution of Ecology to an Understanding of the Natural World.* Oxford: Blackwell Scientific.

Sinclair, A.R.E. and Norton-Griffiths, M. 1979 *Serengeti Dynamics of an Ecosystem.* Chicago: University of Chicago Press.

Smith, E.L. 1984 Use of inventory and monitoring data for range management purposes. In National Research Council/National Academy of Sciences *Developing Strategies for Rangeland Management*. Boulder, CO.: Westview Press.

Snyman, A.H. 1985 Vogbalansstudies op natuurlike veld van die sentrale-Oranje Vrystaat. Ph.D. thesis, University of the Orange Free State, Bloemfontein, Republic of South Africa.

Society for Range Management 1964 A glossary of terms used in range management. Denver, Colorado, USA: Society for Range Management.

Society for Range Management 1974 A glossary of terms used in range management. 2nd edition. Compiled by M.M. Kothmann. Denver, Colorado, USA: Society for Range Management.

Society for Range Management 1989 A glossary of terms used in range management. 3rd edition. Compiled by P.W. Jacoby. Denver, Colorardo, USA: Society for Range Management.

Solomon, K. 1988 *A Study of Range Vegetation at Sarite Ranch in Southern Ethiopia with Particular Emphasis on the Woody Shrubs* Acacia seyal *and* Acacia horrida. Draft Masters of Applied Science thesis, Department of Wool and Animal Science, University of New South Wales, Kensington, Australia.

Soumare, S. 1983 *Les terres salées du Gourma Malien.* Document du Programme 162. Bamako: ILCA (International Livestock Centre for Africa).

Southey, C. 1981 *Land Tenure in Transkei: A report prepared for the Planning Committee.* Mimeo, University of Guelph, Ontario.

Stafford Smith, D.M. 1984 *Behavioural Ecology of Sheep in the Australian Arid Zone.* Ph.D. thesis, Australian National University, Canberra.

Stafford Smith, D.M. 1988 Modeling: three approaches to predicting how herbivore impact is distributed in rangelands. *New Mexican Agriculture Station Regional Research Report* 628. Las Cruces, New Mexico.

Stafford Smith, D.M. 1992a Successors to succession. *Bulletin of the Ecological Society of Australia* 22: 27-30.

Stafford Smith, D. M. 1992b Stocking rate strategies across Australia: Or, how do you cope with drought? *Range Management Newsletter* (Australia) 92(1): 1-2.

Stafford Smith, D.M. and Foran, B.D. 1990 RANGEPACK: The philosophy underlying the development of a microcomputer-based decision support system for pastoral land management. *Journal of Biogeography* 17: 541-546.

Stafford Smith, D.M. and Foran, B.D. 1992 An approach to assessing the economic risk of different drought management tactics on a South Australian pastoral sheep station. *Agricultural Systems* 39: 83-105.

Stafford Smith, D.M. and Morton, S.R. 1990 A framework for the ecology of arid Australia. *Journal of Arid Environments* 18: 255-278.

Stafford Smith, D.M. and Pickup, G. 1990 Pattern and production in arid lands *Proceedings of the Ecological Society of Australia* 16: 195-200.

Standing Committee on Agriculture 1990 *Feeding Standards for Australian Livestock: Ruminants.* Melbourne: CSIRO (Commonwealth Scientific and Industrial Research Organisation).

Stevens, M. and Peggs, A. 1988 ZACK - a computerized farm budgeting program. *Journal of Agriculture of Western Australia* 29: 107-110.

Stocking, M. 1984 *Erosion and Soil Productivity.* First consultants' report, Soil Conservation Programme. Rome: AGLS, FAO (Food and Agriculture Organisation of the United Nations).

Stocking, M. 1985 *Erosion-induced Loss in Soil Productivity: A Research Design.* Consultant's working paper No. 2. Rome: Land and Water Development Division, FAO (Food and Agriculture Organisation of the United Nations).

Stocking, M. and Pain, A. 1984 Soil life and the minimum soil depth for productive yields: developing a new concept. *Discussion Paper* No. 150. Norwich: School of Development Studies, University of East Anglia.

Stocking, M.A. and Peake, L. 1985 *Erosion-induced Loss in Soil Productivity: Trends in Research and International Cooperation.* Norwich: Overseas Development Group, School of Development Studies, University of East Anglia, and Rome: Land and Water Development Division, FAO (Food and Agriculture Organisation of the United Nations). Paper to the IVth International Conference on Soil Conservation, Maracay, Venezuela.

Stoddart, L.A. 1952 Problems in estimating grazing capacity of ranges. In *Proceedings of the 6th International Grassland Congress,* pp. 1367-1373. State College, Pennsylvania.

Stoddart, L.A. 1960 Determining correct stocking rate on range land. *Journal of Range Management* 13: 251-255.

Stoddart, L.A. and Smith, A.D. 1943 *Range Management.* New York: McGraw Hill.

Stoddart, L.A. and Smith, A.D. 1955 *Range Management.* New York: McGraw Hill.

Stoddart, L.A., Smith, A.D. and Box, T.W. 1975 *Range Management.* New York: McGraw Hill.

Stuth, J.W., Conner, J.R., Hamilton, W.T., Riegel, D.A., Lyons, B.G., Myrick, B.R. and Couch, M.J. 1990 RSPM: A resource systems planning model for integrated resource management. *Journal of Biogeography* 17: 531-540.

Sutter, J.W. 1987 Cattle and inequality: herd size differences and pastoral production among the Fulani of North Eastern Senegal. *Africa* 57: 196-218.

Swift, J. (ed.) 1984 *Pastoral Development in Central Niger.* Naimey: USAID (United States Agency for International Development) and Ministère du Développement Rural.

Swift, J. 1990 Savanna ecology and pastoral production: management issues for the future. Closing comments at Commonwealth Secretariat Technical Meeting on Savanna Development and Pasture Production, Woburn, UK, Nov. 1990.

Tainton, N.M. 1981 The ecology of the main grazing lands of South Africa. In Tainton, N. M. (ed.) *Veld and Pasture Management in South Africa.* Pietermaritzburg, South Africa: Shuter & Shooter and University of Natal Press.

Tainton, N.M. 1988 A consideration of veld condition assessment techniques for commercial livestock production in Southern Africa. *Journal of the Grassland Society of Southern Africa* 52: 76-79.

Tamene, Y. 1990 *Population Dynamics of the Problem shrubs, Acacia drepanolobium and Acacia brevispica in the Southern Rangelands of Ethiopia.* Master of Applied Science thesis, Department of Wool and Animal Science, University of New South Wales, Kensington, Australia.

Tapson, D.R. 1982 *Proposals for a Cattle Marketing Strategy for Transkei.* ARDRI Report No 1/82. University of Fort Hare, Republic of South Africa.

Taylor, J.A., Hedges, D.A. and Whalley, R.B.D. 1985 Effects of fertilizer and grazing sheep on pasture heterogeneity in a small-scale grazing experiment. *Australian Journal of Agricultural Research* 36: 315-325.

Taylor, W.J., Neal, J.A. and Rudder, T.H. 1987 'STATIC' and 'HERD': Beef herd management evaluation programs for northern Australia. In White, D.H. and Weber, K.M. (eds.) *Computer-Assisted Management of Agricultural Production Systems*, pp. 102-107. Werribee, Victoria, Australia: Victorian Department of Agriculture and Rural Affairs.

Thornes, J.B. 1990 The interaction of erosional and vegetational dynamics in land degradation: Spatial outcomes. In Thornes, J.B. (ed.) *Vegetation and Erosion*, pp. 41-53. Chichester: John Wiley & Sons.

Thorrington-Smith, E., Rosenberg, M. and McCrystal, L. 1978 *Towards a Plan for KwaZulu: A Preliminary Development Plan*, Vol 1. Ulundi: KwaZulu Government.

Tilman, D. 1988 *Plant Strategies and the Dynamics and Structure of Plant Communities*. Princeton, NJ: Princeton University Press.

Timberlake, J.R. and Reddy, S.J. 1986 *Potential pasture productivity and livestock carrying capacity over Mozambique*. Instituto Nacional de Investigaa Agronmica publication No. 49. Maputo, Mozambique.

Tongway, D.J. 1990 Soil landscape processes in the restoration of rangelands. *Australian Rangelands Journal* 12: 54-57.

Tongway, D.J. and Ludwig, J. 1990 Vegetation and soil patterning in semi-arid mulga lands of eastern Australia. *Australian Journal of Ecology* 15: 23-34.

Trollope, W.S.W., Trollope, L.A. and Bosch, O.H.J. 1990 Veld and pasture management terminology in Southern Africa. *Journal of the Grassland Society of Southern Africa* 71: 52-61.

Tucker, C.J., Dregne, H. and Newcomb, W. 1991 Expansion and contraction of the Sahara Desert from 1980 to 1990. *Science* 253: 299-301.

Tucker, C.J., Justice, C.O. and Prince, S.D. 1986 Monitoring the grasslands of the Sahel 1984-1985. *International Journal of Remote Sensing* 711: 1571-1582.

Tucker, C.J., van Praet, C., Boerwinkle, E. and Gaston, A. 1983 Satellite remote sensing of total dry matter accumulation in the Senegalese Sahel. *Remote Sensing Environ* 13: 461-474.

Turner, J.R. and Tainton, N.M. 1989a Interrelationships between veld condition, herbage mass, stocking rate and animal performance in the tall grassland of Natal. *Journal of the Grassland Society of Southern Africa* 64: 175-182.

Turner, J.R. and Tainton, N.M. 1989b Individual animal performance in relation to herbage mass and stocking rate in the Natal sour sandveld. *Journal of the Grassland Society of Southern Africa* 64: 183-185.

Tyson, P. 1978 Rainfall change over South Africa during the period of meteorological record. In Werger, M. (ed.) *Biogeography and Ecology of Southern Africa*. The Hague: Junk.

USAID (United States Agency for International Development) 1980 *Land Conservation and Range Development Project*. Project Paper, USAID Mission, Maseru, Lesotho.

Vallentine, J.F. 1990 *Grazing Management*. San Diego: Academic Press.

Vanpraet, C.L., Sharman, M.J. and Tucker, C.J. 1983 Utilisation des images NOAA pour l'estimation de la production primaire en pilieu sahélien. In Vanpraet, C.L. (ed.) *Mèthodes d'Inventaire et de Surveillance Continue des Ecosystemes Pastoreaux Sahéliens*, pp. 299-319, Dakar: ISRA/FAO (Food and Agriculture Organisation of the United Nations).

Varley, G. and Gradwell, G. 1960 Key factors in population studies. *Journal of Animal Ecology* 29: 399-401.

Veeneklaas, F.R., Cissé, S., Gosseye, P.A., Duivenbooden, van N., and Keulen, van H. 1990 Compétition pour des ressources limitées: Le cas de la cinquième région du Mali. Rapport 4 : Scenarios de Dévéloppement. Wageningen and Mopti, Mali: CABO (Centre for Agrobiological Research).

Venter, J., Liggitt, N., Tainton, N.M. and Clarke, G.P.Y. 1989 The influence of different land-use practices on soil erosion, herbage production and on grass species richness and diversity. *Journal of the Grassland Society of Southern Africa* 62: 89-98.

Verhulst, P.F. 1838 Notice sur la loi que la population suit dans son accroissement. *Correspondence Mathématique et Physique* 10: 113-121. Translated by L. Wilson. In Kormondy, E.J. (ed.) *Readings in Ecology* pp. 64-66. Englewood Cliffs, NJ.: Prentice-Hall.

Wagenaar, K.T. and de Ridder, N. 1985 *Estimates of Biomass Production and Distribution in ILP Zone in 1985 Based on Satellite NDVI Values.* Working Document. Addis Ababa: ILCA (International Livestock Centre for Africa).

Walker, B.H. 1980 Stable production versus resilience: a grazing management conflict. *Proceedings of the Grassland Society of Southern Africa* 15: 79-83.

Walker, B.H. 1985 Structure and function of savannas: An overview. In Tothill J.C. and Mott, J.J. (eds.) *Ecology and Management of the World's Savannas.* Canberra: Australian Academy of Science.

Walker, B.H., Emslie, R.H., Owen-Smith, R.N. and Scholes, R.J. 1987 To cull or not to cull: Lessons from a southern African drought. *Journal of Applied Ecology* 24: 381-401.

Walker, B.H., Ludwig, D., Holling, C.S. and Peterman, R.M. 1981 Stability of semi-arid savanna grazing systems. *Journal of Ecology* 69: 473-498.

Walker, B., Matthews, D. and Dye, P. 1986 Management of grazing systems - existing versus an event-oriented approach. *South African Journal of Science* 82: 172.

Walker, B.H. and Noy-Meir, I. 1982 Aspects of the stability and resilience of savanna ecosystems. In Huntley, B.J. and Walker, B.H. (eds.).

Walker, J., Raison, R.J. and Khanna, P.K. 1986 Fire. In Russell, J.S. and Isbell, R.F. (eds.) *Australian Soils: The Human Impact,* pp. 185-216. St. Lucia, Australia: University of Queensland Press.

Walter, H. 1971 *Ecology of Tropical and Subtropical Vegetation.* London: Oliver and Boyd.

Warren, A. and Agnew, C. 1988 An assessment of desertification and land degradation in arid and semi-arid areas. *Drylands Programme Research Paper* No. 2. London: IIED (International Institute for Environment and Development).

WCS (World Conservation Strategy) 1980 *The World Conservation Strategy: Living Resource Conservation of Sustained Development.* Glands: IUCN (International Union for the Conservation of Nature and Natural Resources).

Weaver, J.E. and Clements, F.E. 1938 *Plant Ecology.* New York: McGraw Hill.

Webb, P. and Coppock D.L. (In prep.) Effects of pastoral wealth on loss of livestock assets during drought in semi-arid Ethiopia. In preparation for *Human Ecology.*

Weltzin, J.F. and Coughenour, M.B. 1990 Savanna tree influence on understory vegetation and soil nutrients in northwestern Kenya. *Journal of Vegetation Science* 1: 325-334.

Western, D. 1982 The environment and ecology of pastoralism in arid savannas. *Development and Change* 13: 183-211.

Western, D. and Finch, V. 1986 Cattle and pastoralism: survival and production in arid lands. *Human Ecology* 141: 77-94.

Westoby, M. 1980 Elements of a theory of vegetation dynamics in arid rangelands. *Israel Journal of Botany* 28: 169-194.

Westoby, M., Walker, B.H. and Noy-Meir, I. 1989 Opportunistic management for rangelands not at equilibrium. *Journal of Range Management* 42: 266-274.

White, C. 1986 Food shortages and seasonality in WoDaabwe communities in Niger: Seasonality and poverty. *IDS Bulletin* 173: 19-26. Brighton, U.K: Institute of Development Studies.

Whitlow, J. 1980 Environmental constraints and population pressures in the tribal areas of Zimbabwe. *Zimbabwe Agriculture Journal* 77: 173-181.

Whitlow, R. 1987 A national soil erosion survey for Zimbabwe. *Journal of Soil and Water Conservation* 424: 239-242.

Whittaker, R.H. 1951 A criticism of the plant association and climatic climax concepts. *Northwest Science* 25: 17-31.

Wiens, J. 1977 On competition and variable environments. *American Scientist* 65: 590-597.

Wiens, J.A. 1984 On understanding a non-equilibrium world: Myth and reality in community patterns and processes. In Strong, D.R., Simberloff, D., Abele, L.G. and Thistle, A.B. (eds.) *Ecological Communities: Conceptual Issues and the Evidence.* Princeton, N.J.: Princeton University Press.

Wight, J.R. and Siddoway, F.H. 1985 Determinants of soil loss tolerance for rangelands. In *Determinants of Soil Loss Tolerance.* American Society of Agronomy Publication No. 45: 67-74.

Wijngaarden, W. van 1985 *Elephants-Trees-Grass-Grazers: Relationships between Climate, Soils, Vegetation and Large Herbivores in a Semi-arid Ecosystem Tsavo, Kenya.* ITC Publication No. 4. Enschede, Netherlands.

Williams, J.R., Putman, J.W. and Dyke, P.T. 1985 Assessing the effect of soil erosion on productivity with EPIC. In *Erosion and Soil Productivity.* American Society of Agricultural Engineers Publication No. 8/85: 215-225.

Wilson, A.D. and Macleod, N.D. 1991 Overgrazing: present or absent? *Journal of Range Management* 445: 475-482.

Wilson, A.D. and Tupper, G.J. 1982 Concepts and factors applicable to the measurement of range condition. *Journal of Range Management* 356: 684-689.

Wilson, E.O. 1980 *Sociobiology.* Cambridge, Massachusetts: Belknap/Harvard University Press.

Wilson, R.T. 1986 Livestock production in central Mali: Long-term studies on cattle and small ruminants in the agropastoral system. *ILCA Research Report No. 14.* Addis Ababa: ILCA (International Livestock Centre for Africa).

Winrock, 1991 *Interim Report of the Assessment of Animal Agriculture in Sub-Saharan Africa.* Arlington, U.S.A.: Winrock International.

Wischmeier, W.H. and Smith, D.D. 1978 *Predicting Rainfall Erosion Losses, A Guide to Conservation Planning.* Agriculture Handbook No. 537, United States Department of Agriculture. Washington, D.C.: United States Department of Agriculture.

Workman, J.P. 1986 *Range Economics.* New York: Macmillan.

Wylie, B., Harrington, J., Pieper, R., Maman, A. and Denda, I. 1988 *1987 Pasture Assessment Early Warning System.* Integrated Livestock Production Project, Niamey: Government of Niger and Tufts University/USAID (United States Agency for International Development)/New Mexico State University.

Wylie, B., Senock, R., Synder, L., Roettgen, B. and Porter, B. 1983 *Range Research and Results.* NRL preliminary report. Niamey: USAID (United States Agency for International Development).